SUBMARINE MINES

AND

TORPEDOES

AS APPLIED TO HARBOUR DEFENCE.

SUBMARINE MINES

AND

TORPEDOES

AS APPLIED TO HARBOUR DEFENCE

JOHN TOWNSEND BUCKNILL

HONORARY LIEUTENANT-COLONEL
(LATE MAJOR R.E.), RESERVE OF OFFICERS

The Naval & Military Press Ltd

published in association with

FIREPOWER
The Royal Artillery Museum
Woolwich

Published by
The Naval & Military Press Ltd
Unit 10 Ridgewood Industrial Park,
Uckfield, East Sussex,
TN22 5QE England
Tel: +44 (0) 1825 749494
Fax: +44 (0) 1825 765701
www.naval-military-press.com

in association with

FIREPOWER
The Royal Artillery Museum, Woolwich
www.firepower.org.uk

PREFACE.

IN this book-producing age, the man who writes one owes an apology to the public.

The following pages have, for the most part, already appeared in *Engineering*, in a series of articles on submarine mining, and several requests having been received to republish same in book form, it has been determined to do so.

An examination of the Table of Contents enables a reader to separate theory from practice. The general sequence of the chapters has been arranged as far as possible in the order in which a scientific subject should be investigated, viz.: (1) Theory; (2) Experiment; (3) Practical Application. In compiling these pages, great care has been taken to guard all Government secrets —those arrangements only being described which have already been published, either in the specifications obtainable at the Patent Office, or in the public press, or in books sold to the public in this and other countries. I shall feel compensated for a long year of labour if the general ideas propounded receive attention, and should they be accepted, I shall feel that it is more due to the exceptional advantage of having worked for several years in daily contact with the English Vauban than to any perspicuity of my own. The broad principles of scattering submarine mines as much as possible should be identified with him (Sir Wm. F. Drummond Jervois), and if he agree in the main with the contents of the following pages I shall consider their accuracy proven.

The three concluding chapters on Torpedoes will enable the reader to form an opinion as to the value of these weapons for harbour defence, and as to the manner in which they can best be employed in conjunction with other arms. General Abbot's excellent work on "Experiments with Submarine Mines in the United States" has been of the greatest assistance, although I have taken exception to some of his deductions. Many other authorities have been quoted and are duly acknowledged in the text. My own inventions connected with submarine mining are perhaps noticed too prominently, but it was necessary to illustrate the various and necessary contrivances by some special patterns, and it not unfrequently occurred that I was compelled to describe my own patterns or none at all, the Government gear being treated as secret and confidential.

For a similar reason, it being impossible to indicate the defects of our service arrangements with precision, I have been compelled to attack their intricacy in general terms only.

On matters connected with the *personnel* and the purchase of stores, however, it has been possible to go into detail; and if some of the remarks appear to be harsh, my excuse must be that the subjects required it. Minced language is not a desirable form of expression when the writer believes that the efficiency of an important item in National Defence is at stake.

J. T. B.

Southampton, Dec., 1888.

CONTENTS.

LIST OF DIAGRAMS.

LIST OF TABLES.

FORMULÆ.

SUBMARINE MINING.

CHAPTER I.—INTRODUCTORY.

Also Analysis of Important Experiments, and of Actual Results in War.

Submarine warfare, whether it be carried on by means of mines or torpedoes, or shells fired through the air from a distance, depends upon the fact that the explosion under water of a charge of proper dimensions properly placed, will damage the vessel attacked so as to place her *hors de combat*, or destroy her. The idea of attacking vessels in this manner must have suggested itself to many minds long before we have any records of actual attempts being made, but the difficulties of placing a charge and of firing it with certainty when so placed, were enormous only a hundred years since; and, what now appears so simple a matter, owing to the progress in science, must then have been considered an almost impossible problem to any but sanguine inventors.

Nevertheless, more than a century has elapsed since British ships were subjected to the attack of drifting torpedoes in the Delaware River during the War of Independence; and early in the present century both Fulton and Warner endeavoured to persuade European nations to adopt their ideas, but without success. Those were hard times for inventors.

More recently the Russians used small gunpowder charged mines in the Baltic during the Crimean War, but the chemical fuze employed was slow in its action, and the results were insignificant.

The birth of the submarine mine and of the torpedo in practical forms occurred in the American War of Secession, and it will be interesting to record the damage done by these weapons during that war. The results obtained are astounding, for, at the commencement of the war, the Confederates possessed no special stores, no trained personnel, and but little scientific knowledge of the subject:—

1. In December, 1862, the U.S.N. armoured vessel Cairo, 512 tons, 13 guns, was destroyed by a mine in the Yazoo River.

2. In February, 1863, the U.S.N. monitor Montank, 844 tons, 2 guns, was seriously injured by a mine in the Ogeechee River.

3. In July, 1863, the U.S.N. armoured vessel Baron de Kalb, 512 tons, 13 guns, was destroyed by a mine in the Yazoo River.

4. In August, 1863, the U.S.N. gunboat Commodore Barney, 513 tons, 4 guns, was disabled by an electrical observation mine in the James River. Charge, 2000 lb. gunpowder. Ignition rather late.

5. In September, 1863, the U.S.A. transport John Farron was seriously injured by a mine in the James River.

6. In October, 1863, the U.S.N. armoured vessel Ironsides, 3486 tons, 18 guns, was seriously injured by a spar torpedo boat off Charlestown. Charge, 60 lb. gunpowder.

7. In 1863 the Confederate vessel Marion was destroyed by a mine accidentally when laying mines off Charlestown.

8. In 1863 the Confederate vessel Eltiwan was seriously injured by a mine off Charlestown.

9. In February, 1864, the U.S.N. sloop of war Hoosatonic, 1240 tons, 13 guns, was destroyed by a spar torpedo boat off Charlestown.

10. The torpedo boat itself was sunk and was never seen or heard of again.

11. In April, 1864, the U.S.A. transport Maple Leaf, 508 tons, was destroyed by a mine in the St. John's River.

12. In April, 1864, the U.S.A. transport General Hunter, 460 tons, was similarly destroyed.

13. In April, 1864, the U.S.N. flagship Minnesota, 3307 tons, 52 guns, was damaged internally by a spar torpedo boat in Newport News. Charge 53 lb. gunpowder. Submerged 6 ft.

14. In April, 1864, the U.S.N. armoured vessel Eastport, 800 tons, 8 guns, was sunk by a mine in Red River.

15. In May, 1864, the U.S.N. gunboat Commodore Jones, 542 tons, 6 guns, was destroyed by an electrical observation mine in James River. Charge, 2000 lb. gunpowder.

16. In May, 1864, the U.S.A. transport H. A. Weed, 200 tons, was destroyed by a mine in St. John's River.

17. In June, 1864, the U.S.A. transport Alice Price, 320 tons, was destroyed by a mine in the St. John's River.

18. In August, 1864, the U.S.N. monitor Tecumseh, 1034 tons, 2 guns, was destroyed by a mine in Mobile Bay.

19. In October, 1864, the Confederate armoured vessel Albemarle, 2 guns, was destroyed by a spar torpedo boat at Plymouth.

20. The torpedo boat sank.

21. In November, 1864, the U.S.A. transport Greyhound, 900 tons, was destroyed by a coal mine in her furnace in James River.

22. In December, 1864, the U.S.N. gunboat Narcissus, 101 tons, 2 guns, was destroyed by a mine in Mobile Bay.

23. In December, 1864, the U.S.N. gunboat Otsego, 974 tons, 10 guns, was destroyed by a mine in the Roanoke River.

24. In December, 1864, the U.S.N. tug Bazley was destroyed by a mine in the Roanoke River.

25. In January, 1865, the U.S.N. monitor Patapsco, 844 tons, 2 guns, was destroyed by a mine off Charlestown.

26. In February, 1865, the U.S.N. gunboat Osceola, 974 tons, 10 guns, was crippled by a drifting torpedo in Cape Fear River.

27. In 1865, the Confederate transport Shultz was destroyed by a mine accidentally in the James River.

28. In March, 1865, the U.S.N. gunboat Harvest Moon, 546 tons, 3 guns, was destroyed by a mine at Charlestown.

29. In March, 1865, the U.S.A. transport Thorne, 403 tons, was destroyed by a mine in James River.

30. In March, 1865, the U.S.N. gunboat Althea, 72 tons, 1 gun, was destroyed by a mine in Blakely River.

31. In March, 1865, the U.S.N. monitor Milwaukee, 970 tons, 4 guns, was destroyed by a mine in Blakely River.

32. In March, 1865, the U.S.N. monitor Osage, 523 tons, 2 guns, was destroyed by a drifting torpedo in the Blakely River.

33. In April, 1865, the U.S.N. gunboat Rodolph, 217 tons, 6 guns, was destroyed by a mine in Blakely River.

34. In April, 1865, the U.S.N. gunboat Ida, 104 tons, 1 gun, was destroyed by a mine in Blakely River.

35. In April, 1865, the U.S.N. gunboat Sciota, 507 tons, 5 guns, was destroyed by a mine in Mobile Bay.

36. In May, 1865, the U.S.A. transport R. B. Hamilton, 400 tons, was destroyed by a mine in Mobile Bay.

37. In June, 1865, the U.S.N. gunboat Jonquil, 90 tons, 2 guns, was seriously injured by a mine when raising frame torpedoes in Ashley River.

When the employment of mines and torpedoes was first commenced by the Confederates, the Northerners affected to treat them with indifference. This feeling gradually wore away. The long list of vessels destroyed proved the efficiency of submarine mines more thoroughly than any amount of argument.

A certain number of people, especially those interested in gunnery, and more recently those connected with torpedo boats or with locomotive torpedoes, assert that the sphere of action of a submarine mine is very limited. But they forget that the position of a mine being unknown to a foe, the whole of any waters which may be mined must be treated as if they are known to be mined. To insure this end, mines should be scattered as much as possible, and the greatest secrecy should be maintained concerning their intended positions, the plans of the mine fields, and the approximate position of the mine fields being known only to a selected few, the number of the mines to be used in any harbour being kept secret, and misleading reports spread concerning all these matters.

Soon after the American War of Secession, European nations took up the subject in earnest, trained men in the preparation and planting of mines, and purchased the necessary stores and appliances. Committees were formed to investigate and report upon the matter, after making the necessary experiments; but it was not until nearly ten years afterwards that experiments with targets representing the hull of a modern warship were made in Europe to discover with exactitude the distances of destructive effect of various submarine explosions.

England led the way by the long and important series of experiments against the hull of H.M.S. Oberon, which was altered so as to represent the bottom of the strongest ironclad then afloat, viz., H.M.S. Hercules, which has an outer skin about $\frac{7}{8}$ in. thick supported by frames forming rectangles about 6 ft. by 4 ft., and 3 ft. to the inner skin.

These experiments were described somewhat minutely in the *Times* by their able reporter at Portsmouth, and foreign governments thus obtained much useful information.

General Abbot, of the U.S. Engineers, has for several years been engaged in the investigation of the effects of submarine explosions, and his report to Congress on the subject is a classic, and contains tabulated information concerning the Oberon experiments, but the pressures recorded in the column marked P (Table I.), and which are calculated from General Abbot's formula to be examined in a following chapter, may be incorrect, as the formula does not give results agreeing with some other important and carefully conducted experiments. If corrrect, "a study of the figures in Table I., and of the injuries inflicted, leads to the conclusion that an instantaneous mean pressure of 5500 lb. per square inch exceeded the resisting power of the Oberon; and hence that such a blow would cripple the Hercules in action."

A large number of crusher gauges were attached to the sides of the Oberon, and the results as recorded on them have never been published. They were unsatisfactory, probably due to water getting into the gauges.

TABLE I.—ABBOT'S ANALYSIS OF THE "OBERON" EXPERIMENTS, 1874-76.

No. of experiment.	Explosive.		Distance.		Angle from Nadir (a).	Submersion of Charge.	Depth of Water in feet.		Pressure (P) in Pounds per sq. in. on nearest point of Hull. Abbot's Formula.	Ditto (P') Calculated by a New Formula now proposed by Author.	REMARKS. The last Column has been added to General Abbot's Table.
	Description.	Weight (C).	Actual (D).	Horizontal.			At Charge.	At Oberon.			
		lb.	ft.	ft.	deg.	ft.					
1	Wet G.C. (disc)	500	109	100	113	48	48	60	1,235	4,348	Hull shaken. Condenser pipe split. No serious damage.
2	,,	500	91	80	118	48	48	60	1,609	5,269	Hull shaken. No rupture.
3	,,	500	74	60	125	48	48	60	2,196	6,581	Seriously shaken. No rupture of bottom. Sea connections damaged.
4	,,	500	66	50	131	48	48	60	2,612	7,467	Outer plating buckled. Rivets started. No leak. Condenser, &c., seriously damaged.
5	,,	500	52	30	137	48	66	70	3,697	9,644	Outer plating much buckled. No leak.
6	,,	500	52	30	137	48	48	48	3,697	9,644	Small leaks started. No fatal rupture. Outer skin seriously damaged.
7	,,	500	38½	0	164	48	48	48	5,996	15,918	Fatal shock. Ship sank. Much damage of various kinds.
8	Wet G.C. (slab)	60	15	15	100	10	30	30	4,927	4,085 (5,106)	Outer skin indented 1½ in. (This experiment is erroneously recorded C=75, and P is therefore too small.)
9	Gunpowder	66	3	3	100	9	30	30	4,155	19,541	Fatal local shock. Large hole opened through both skins.
10	Wet G.C. (slab)	33	4	4	100	9	30	30	19,800	19,430	Ditto ditto.
11	Do. granulated	33	4	4	100	9	30	30	19,800	19,430	Ditto ditto.

TABLE II.—Abbot's Analysis of the Carlskrona Experiments.

No. of Experiment.		Explosive. Description.	Submergence (S).	Charge (C).	Distance (D).		Angle from Nadir (α).	Pressure (P) in Pounds per sq. in. on nearest point of Hull. Abbot's Formula.	Ditto (P') Calculated by a New Formula now proposed by Author.	REMARKS.
			ft.	lb.	ft.	in.	deg.	lb.		
First series.	1	Dynamite	7	13	2	2	90	27,865	32,856	Wooden side. Hole 15 ft. × 8 ft. through timbers.
	2	,,	7.7	16	3	0	90	20,300	18,240	,, ,, ,, 8 ,, × 8 ,, ,, ,,
	3	,,	5.7	16	2	0	90	35,765	52,200	,, ,, ,, 10½ ,, × 9 ,, ,, ,,
	4	,,	6.5	10	2	2	90	23,395	24,920	,, ,, ,, 4 ,, × 16 ,, ,, ,,
	5	,,	7.3	13	2	2	90	27,865	32,400	Iron ,, ,, 4 ,, × 3 ,, ,, plates.
Second series.	1	,,	9.2	33	25	6	90	1,652	1,210	No serious damage.
	2	,,	9.2	47.2	25	6	90	2,114	1,731	Leak from loosened rivets.
	3	Rifle powder	9.2	112	12	0	90	1,610	2,463	Ship shaken severely. Many rivets started.
	4	Dynamite	9.2	33	15	0	90	3,478	2,205	No serious damage.
	5	,,	9.2	66	21	0	90	3,442	4,526	Ditto.
	6	,,	9.2	33	12	9	90	4,358	2,678	Severe shock to vessel. No serious injury.
	7	,,	9.2	33	4	6	90	22,040	14,718	Hole 14 × 12. Both bottoms ruptured.
	8	Gunpowder	29.2	660	32	3	130	3,813	5,245	No injury to vessel.
	9	Dynamite	9.2	19	10	6	90	3,958	2,003	Ship shaken severely. Plates indented.
	10	,,	9.2	19	3	3	90	19,950	17,887	Hole 6½ × 2 to 5. Inner skin bulged and crushed.
	11	Gunpowder	9.2	112	5	9	90	3,857	7,669	,, through both bottoms.
	12	,,	29	660	23	9	160	6,114	8,274	,, 100 square feet in outer and 75 square feet in inner skin of bottom. Ship destroyed.

An examination of experiments 9 and 10 on Table I. shows that there must be something wrong with the pressures calculated from Abbot's equation, as the smaller pressure could not possibly produce so much the greater effect. Another equation proposed by the author, and to be examined in a future chapter, gives better results, which are shown in the last column of the Table now being examined. If correct, the pressure required to produce a fatal effect on an ironclad is much nearer 12,000 lb. on square inch than 5500 lb.

The values for intensity of action (I) in calculations for P^1 were taken at 25 for gunpowder and 100 for gun-cotton. In the column for P the value 87 was given to gun-cotton, which is probably 13 per cent. too low. The values for D are only correct approximately, the precise form of the Oberon's hull being unknown to General Abbot. The weight of the hull, &c., was 1100 tons; draught, 11 ft.; length, 164 ft.; beam, 28 ft. 6 in.

A condenser was fitted to the Oberon, but no boiler or machinery. A sheep and other animals on board were not injured by any of the experiments. The author remained on board during one of the experiments not recorded in the Table (100 lb. gun-cotton 25 ft. off). The effect was a sharp jar on the ankle bones. Many other experiments were made with the Oberon, but the more important are all recorded in the Table now borrowed from General Abbot.

About the same date another series of important experiments was carried out at Carlskrona by a combined committee of Danish and Swedish officers. An iron target representing the side and bottom of H.M.S. Hercules was inserted in a very strong wooden ship named the Vorsicticheten.

These experiments were published in Commander Sleeman's book on "Torpedoes, 1880," and Abbot's analysis now reproduced in Table II. was drawn up from it. A column is now added for the pressure P^1 calculated by the author's formula to be described hereafter. An examination of the Table indicates that the results are better explained by column marked P^1 than by column marked P.

If experiments numbered 8, 9, 11 be compared, the pressures under P are nearly the same, although the results are so widely different. Thus "no injury" is recorded against No. 8; "plates indented" against No. 9; and "a hole through both bottoms" against No. 11.

Some experiments were carried out by the Austrian Government at Pola in 1875, and are recorded in Abbot's report. Target, a pontoon; draught of water, 19 ft.; 60 ft. long, 40 ft. beam, with circular ends, and fitted with a condenser and two Kingston valves, and with a double bottom to represent the Hercules.

TABLE III.—ABBOT'S ANALYSIS OF MISCELLANEOUS EXPERIMENTS, &c.

Number of Experiment.	Target.	Explosive. Description.	Explosive. Submersion.	Explosive. Weight.	Distance.	Angle from Nadir. (*a*).	Pressure (P) in Pounds per Square Inch on Nearest Point of Hull. Abbot's Formula.	Ditto (P') Calculated by a New Formula now proposed by Author.	REMARKS.
			ft.	lb.	ft.		lb.		
1	Dorothea	Gunpowder	15	180	3	180	16,020	68,850	Total destruction.
2	Terpsichore ..	Rifle powder	22	150	19	175	1,403	2,527	Hole 8 ft. in diameter.
3	English target ..	Gunpowder	7.5	100	1	90	18,385	maximum	Total destruction.
4	Marie	Gun-cotton	26.2	276	35.6	132	4,172	7,777	Leak, but no rupture.
5	,, ...	,,	39.4	917	54.9	132	5,062	16,551	Strained. Sank slowly.
6	,, ...	,,	75	1554	65.8	165	6,144	25,516	Stove in.
7	Express	,,	27.2	739	36.1	132	7,892	20,526	Destroyed.
8	Eldorado	,,	65.9	1623	65.9	152	6,033	25,366	Crushed; broken in two amidships
9	Marie	Gunpowder	23	441	22	..	3,330	4,743	Unhurt.
10	,, ...	,,	26.2	1102	36	..	6,150	6,383	Insignificant effect.
11	Express	,,	23	441	22	..	3,330	4,743	Small leak.
12	Wagram	,,	19.7	441	15	..	5,139	7,350	Dangerous leak.
13	,, ...	,,	28.2	1323	38	..	7,015	7,969	Leak. No rupture.
14	,, ...	,,	33.7	2205	46	..	10,090	10,910	Ditto.
15	Cormoran	,,	32.8	1102	36	..	6,207	6,383	Hole 18 ft. by 13 ft.
16	Requin	,,	59.3	2205	60	..	7,087	8,326	Hull intact. Rails carried away.
17	,, ...	,,	52.5	3310	53	..	13,550	14,122	Cut in two.
18	Fulton	,,	131	4409	122	..	6,456	8,145	Crushed. Hole 20 ft. long.
19	Gunboat	,,	19.4	450	23	..	3,729	4,610	Small hole 3 sq. ft.
20	,, ...	,,	15.4	287	18	..	2,948	3,864	Destroyed.
21	Audacious	,,	28.9	661	24	..	5,480	6,466	Eleven frames broken. Bulwarks destroyed. No hole.
22	,, ...	,,	40.3	661	33	..	3,677	4,610	Keel broken. Hole 84 sq. ft.
23	Prudence	,,	29.2	655	23	..	5,937	6,710	Holes through both bottoms 92 sq. ft. and 78 sq. ft.
24	,, ...	,,	10.5	112	4	..	5,136	16,144	Ditto, 70 sq. ft. and 57 sq. ft.
25	Austrian sloop ..	Gun-cotton	10	400	24	90	8,285	15,625	Total destruction.
26	,, pontoon ..	Dynamite	40.5	617	53	130	4,576	11,510	Moved 13 ft. bodily.
27	,, ,, ..	,,	36	585	48	130	5,075	12,073	Moved 4 ft. Shaken.
28	Large gunboat ..	Gunpowder	25 (?)	200	15	180	2,463	4,500	Complete destruction.
29	Turkish schooner ..	Gun-cotton	10	100	1	90	377,450	..	Ditto.
30	Ironsides	Gunpowder	10 (?)	60	3 (?)	90	3,710	maximum	Severe shock. This charge must have been further off than 3 ft., *vide* Experiment 9, Oberon Table.
31	Minnesota	,,	10	53	3 (?)	90	3,205	..	Ditto.
32	Commodore Jones	,,	36	2000	30 (?)	180	16,333	20,812	Total destruction.
33	Conklin	Mortar powder	13	188	8	180	1,869	9,927	Ditto.
34	Olive Branch ..	,,	7	50+50	3	180	1,421	19,125	Ditto. As 66 lb. similarly situated broke through both skins of Oberon, this result must have been evident.

Experiment 1.—C = 617 lb. dynamite, D = 53 ft., D horizontal = 62 ft. from keel, submersion 40½ ft.; depth of water 60½ ft. at charge and 62 ft. at target. Effects: outer skin slightly indented, a few rivets started, several screws of valves loosened.

Experiment 2.—C = 585 lb. dynamite, D = 48 ft., D horizontal = 60 ft. from keel, submersion 36 ft.; depth of water, 78 ft. at charge, 74 ft. at target. Effects: some rivets loosened, a few angle irons sheared, outer skin slightly indented, no damage to condenser or valves. In each experiment the charge was placed opposite the centre of the pontoon.

On Table III. is given Abbot's analysis of miscellaneous experiments, &c., a column being added, as before, showing the pressures calculated by the author's formula to be described hereafter.

General Abbot's remarks on this Table are as follows:

1. "Except in the uncertain and anomalous gunpowder trials upon the Wagram and the Requin (the former vessel not described, and the latter a very small craft) in every instance where the computed mean pressure exceeded our adopted standard of 6500 lb. per square inch" (as calculated by General Abbot's equations) "the vessel was destroyed."

2. "The injuries to the Terpsichore, and to the Conklin and Olive Branch, show that an ordinary wooden hull will not always endure a computed mean pressure of 1500 lb. per square inch."

By the author's formula these pressures appear to be much greater than 1500 lb., viz., 10,000 lb. and 19,000 lb.

3. "The injuries inflicted upon the Ironsides and Minnesota indicate that 3000 lb. per square inch exceeds the limit of safety even for a strong wooden hull. The blow received by the latter is especially interesting as fixing her extreme endurance."

Experiment No. 9 on the Oberon series proves conclusively that the distances recorded against the Ironsides and Minnesota must be erroneous.

4. "The trials upon the Austrian pontoon confirm the conclusion that the recoil of a target reduces the effect upon it. . . . . ."

5. "The discrepancies exhibited by some of the gunpowder experiments, and the remarkable uniformity shown by those with gun-cotton and dynamite, confirm the conclusion reached when testing these explosives in rings" fitted with crusher gauges, to be described in another chapter. The conclusion referred to was that explosive mixtures appear to act with a decided maximum intensity in some one direction at each explosion. In other words, that there is a marked burst of gas from such an explosion, but that it is as likely to occur in one direction as another. With explosive compounds no such erratic action

was observed. This is one of several reasons for preferring the compounds to the mixtures for submarine mining.

The following additional information concerning some of the entries on Table III. is taken from General Abbot's report:

1. The Dorothea, a strong 200-ton brig, was blown up in England by Robert Fulton in 1805.

2. H.M.S. Terpsichore, sloop-of-war, 10 ft. draught, was blown up in 1865 in England. The charge was 2 ft. clear of the ship, horizontally.

3. This target was an iron case 20 ft. long by 10 ft. high by 8 ft. deep or thick. It had six compartments, being divided by one $\frac{1}{4}$-in. longitudinal bulkhead and two $\frac{3}{8}$-in. cross bulkheads. The front and rear faces were $\frac{11}{16}$ in. thick.

Large pieces of iron were hurled nearly 200 ft. in the air and to a distance of about 100 yards.

Experiments 4 to 24 on the Table were carried out in France, and but little is known about them. The Prudence had some kind of plating on her side presented to the explosions. The dimensions of the vessels were as follows:

TABLE IV.

	ft.		ft.		ft.	
Requin	95	by	16	by	4.5	draught
Express	108	,,	20	,,	6.6	,,
Marie	115	,,	21	,,	7.2	,,
Cormoran	131	,,	25	,,	7.7	,,
Eldorado	213	,,	39	,,	10.8	,,
Prudence	117	,,	49	,,	14.3	,,

It is impossible to say much about these experiments, because the relative strengths of the hulls and the weights on board and the method of mooring adopted are all unknown.

As a result of the series, the French commission recommended the following charges for ground mines:

TABLE V.

Depth of Water. ft.	Gun-cotton. lb.	Gunpowder. lb.
26 to 36	550	2200
50	660	3300
60	880	4400
67	1100	
73	1320	
80	1540	

The charges of gun-cotton appear to vary nearly as the depths. Thus,

30 : 80 :: 550 : 1467.

At the smallest submersion the gunpowder charge is four times the gun-cotton charge, and the ratio of charge to depth is also retained. Thus,

30 : 60 :: 2200 : 4400.

The charges are certainly larger than necessary, and the practice of employing ground mines in very deep water is not to be recommended.

The Austrian experiments Nos. 26 and 27, on Table III., have already been described. No. 25 was carried out in order to test the effect of a large charge on a wooden vessel. It would, in all probability, have been equally effective at double the distance.

The German gunboat, destroyed in experiment No. 28, Table III., was strengthened internally. The charge was placed directly under the vessel's keel, and the result proves how tremendously the effect on a vessel is increased when the line of least resistance from a gunpowder charge lies through the hull. The force of the explosion is directed, and the damage is much greater than the calculated pressures would indicate. Experiments numbered 29, 30, and 31 on Table III. are of no value; No. 29 because the result ought to have been a foregone conclusion, and Nos. 30 and 31 because the distances of the charges from the hulls are evidently erroneous, which can be proved by comparing with No. 9 (on Table I.) of the Oberon experiments.

The last two experiments on Table III. are chiefly interesting because the pressures as calculated by Abbot's equations are so low, although the vessels were blown to atoms. In the case of the Olive Branch, 50 lb. of gunpowder at 3 ft. gives a calculated pressure of only 1421 lb. on the square inch, according to General Abbot; but in No. 9 experiment on Oberon Table, 66 lb. of gunpowder at 3 ft. he calculated to give a pressure of 4155 lb. on the square inch. The pressures calculated by the formula proposed by the writer of this paper are 19,541 for the one and 19,125 for the other, which appear much more reasonable, and would account for the great destruction recorded.

The formulæ are as follows, and they will be carefully examined in Chapter III.

General Abbot's formula,

$$P = \left(\frac{M\, S^{.025}\, C^{1.94}}{(D+1)^2\, R^{.5}}\right)^{\frac{2}{3}} \text{ for explosive mixtures.}$$

$$P = \left(\frac{6636\,(a+E)\,C}{(D+0.01)^2}\right)^{\frac{2}{3}} \text{ for explosive compounds.}$$

Lieut.-Colonel Bucknill's formula,

$$P = \frac{9\,C\,I}{D}\left(1+\frac{25}{D^2}\right)\left(1+\frac{\beta}{90}\times\frac{e}{100}\right) \text{ for all explosives}$$

and

$$P = \frac{9\,C\,I}{D}\left(1+\frac{25}{D^2}\right) \text{ ditto in horizontal plane only.}$$

In these equations,

P = pressure per square inch on nearest point of target in pounds.
C = charge of explosive in pounds.
D = distance from centre of C to target in feet.

Also :

M = a value found by experiment with the different explosive mixtures.

E = ditto, with the different explosive compounds.

I = relative intensity of action of the explosive, dynamite being 100.

α = angle from nadir of direction of the target from centre of charge, in degrees.

β = angle from horizontal plane of ditto, plus if above, minus if below.

e = a value (in the nature of a percentage) dependent on the explosive.

S = submersion of charge in feet.

R = radius in feet of sphere of explosive mixture ignited by one fuze.

The value of M adopted after experiments by General Abbot	= 790	for mortar powder.
	= 1554	,, musket ,,
	= 3331	,, sporting ,,
The value of E which depends on relative intensity of action of the explosive	= 186	,, dynamite No. 1 (I being 100).
	= 135	,, gun cotton. (I being taken = 87).
The value of I adopted by the author differs somewhat	= 100	,, dynamite No. 1.
	= 100	,, gun-cotton.
	= 25	,, gunpowder.
The value of e for vertical action in the author's formula	= 20	,, dynamite.
	= 20	,, gun-cotton.
	= 35	,, gunpowder.

General Abbot assumes that an instantaneous mean pressure of 6500 lb. on the square inch will give a fatal blow to a modern ironclad.

The author assumes that a pressure of 12,000 lb. is required.

It will now be convenient to examine the apparatus for measuring these pressures, which have been used in various countries.

CHAPTER II.

On the Apparatus Employed for Measuring the Effects of Submarine Explosions.

The records of the early experiments made in order to discover the laws that govern subaqueous explosions are neither interesting nor instructive, except to prove how little was known of the subject twenty years ago. After trying mechanical contrivances of various kinds, the happy thought of examining the surface of a mud flat after the explosion of a charge upon it, suggested itself to a member of one of the English committees, and a series of experiments giving for the first time good and trustworthy results were accordingly instituted. A site was chosen where there was a considerable rise and fall of tide, so that the charge could be placed on the mud at low tide and be fired with a good submergence at high water.

A series of experiments were also made in England, in which the instantaneous photographs of the columns of water thrown up by various charges were examined, and some useful formulæ deduced therefrom by Captain W. de W. Abney, R.E., F.R.S., &c. These formulæ have only been published confidentially, so that they cannot be reproduced here.

The Americans also made some experiments by means of photography, but discarded them as being of insufficient exactitude.

In 1851 General Rodman, of the United States army, invented the pressure gauge known by his name, and used in ordnance experiments. It consisted of a small cylinder containing a piston which drove a V-shaped indenting tool upon a disc of pure copper.

In 1865 Major King, of the United States Engineers, applied it for measuring the effects of subaqueous explosions.

In 1869 Captain W. H. Noble, R.A., invented his crusher gauge for measuring the gas pressures produced inside the bores of guns when they are fired. It consisted of a piston and cylinder, the movement of the former crushing longitudinally a small solid cylinder or pellet of copper, the amount of compression giving a record of the pressure to which the crusher gauge had been subjected, the amount of pellet compression produced by different statical pressures having previously been ascertained experimentally.

Soon afterwards a modification of this crusher gauge was introduced into America for employment in the submarine experiments. Lead pellets were used instead of copper; the inside of the cylinder was roughened by a number of horizontal corrugations, and small spring catches engaging in these corrugations were attached to the piston so that it should not hammer the pellet by a series of blows, experience having shown that this occurred when unroughened gauges were subjected to the effects of submarine explosions. Water was excluded by a rubber cap over the end of the cylinder and secured to it by a band engaging at the groove A (see Figs. 1 to 4). The cylinder was screwed

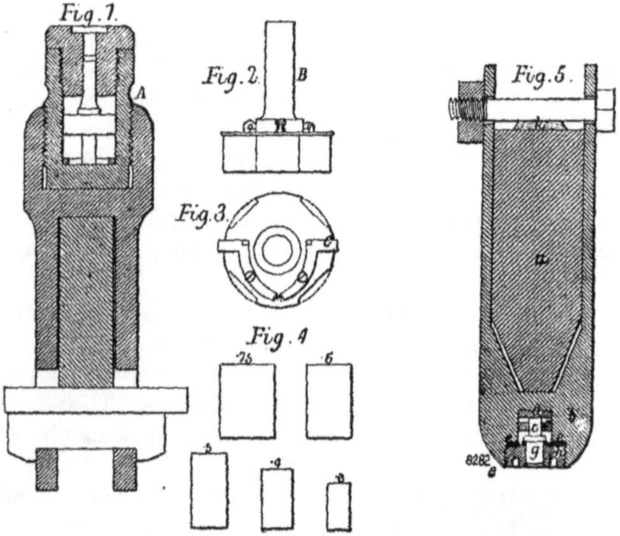

into a socket with two ears to engage an iron ring (shown in section on the drawing), and was fixed thereto by a wedge and cotter.

Pellets of different diameters were used in order to obtain the desired sensitivity. Also rings of different diameters, 3 ft., 4 ft., 5 ft., 6 ft., 8 ft. in diameter. The gauges could also be attached to the iron bars of a large framework crate which was made for the American experiments. This crate was 50 ft. long, the effects as recorded on crusher gauges attached thereto could therefore be obtained up to a distance of about 25 ft. from a charge exploded in the centre.

In 1873 the War Office Torpedo Committee (England) caused similar apparatus to be manufactured, but no provision was made against the hammering action already alluded to, and water was kept out simply by the perfection of the fit between the piston and the cylinder in the crusher gauges. The results of the crusher gauge experiments in England were, on the whole, unsatisfactory, and this may have been due to the omission of the spring catches so carefully fitted to the American gauges. Moreover, when the English crusher gauges were submerged a long time before an experiment was carried out, inaccuracy may have been caused by the entrance of water into the gauge cylinders.

The pellets used in England were 0.5 in. long and 0.326 in. in diameter $= \frac{1}{12}$ square inch sectional area. Both lead, hardened with antimony, and copper were employed. The latter was found to give the most reliable results.

Fig. 6.

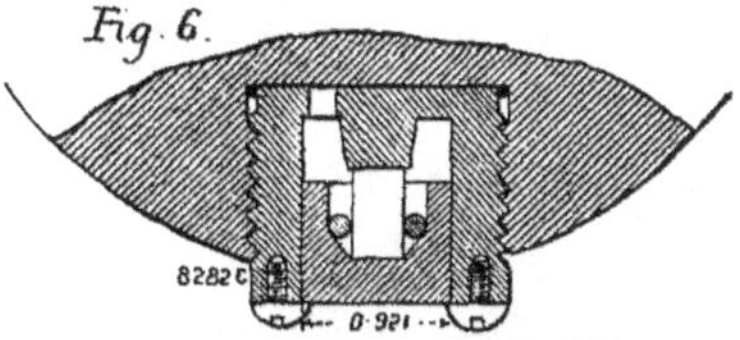

Fig 7.

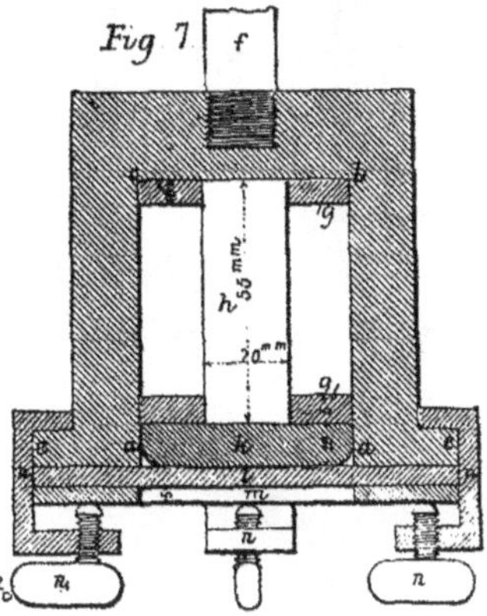

The piston area struck by the explosion varied; in some gauges it was $\frac{1}{6}$ square inch, in others it was as much as $\frac{2}{3}$ square inch.

Fig. 5 shows the general arrangement of one of the 5-ft. ring gauges. *a* is the section of ring; *b* is the socket; *c* lead or copper pellet; *d* steel footplate; *e* rubber washer; *f* rubber ring; *g* steel piston; *h* screw plug and guide for the piston; *k* wooden wedge. A small bent steel spring engages under *h* and over the enlarged portion of the piston, keeping the latter firmly against the pellet.

Another form of English crusher gauge is shown on Fig. 6, and is screwed into the bottom of a solid 18-pounder shot provided with an eye-bolt at the top by which it is suspended from a float. Or the gauge may be one of several screwed into the side of a 13-in. shell, or into the side of a cast-iron sinker, or any other substantial metal body. The pellet is centered by a rubber ring inside a piston which is kept in place by two screws as shown. The pellet is seated on a projection forming part of the steel cylinder. Three holes are provided at the bottom of the cylinder so that the piston can be forced out after the experiment

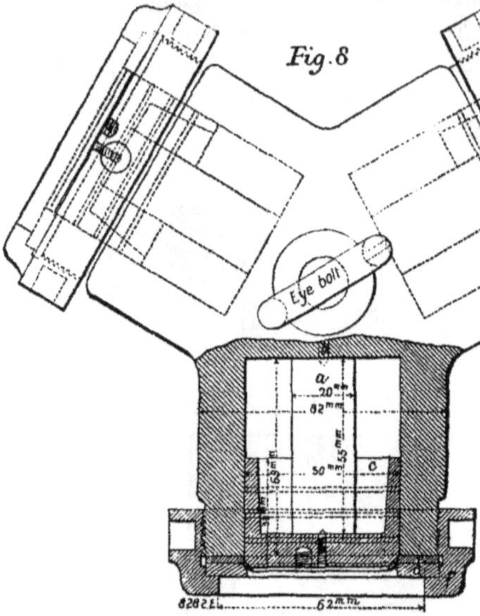

FIG. 9

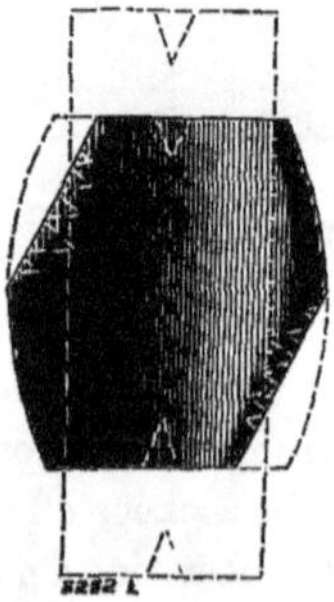

by means of a three-pronged fork. The bottom is made water-tight by a sheet of lead or india rubber. Other modifications were used, and the above are typical of them all.

The results of the English experiments have not been published, but it may be stated that the best as well as the highest compressions for given distances were obtained in the gauges fixed to shot and shell simply suspended in the water and almost free to move with it. This is curiously in contradiction with the published results of the American experiments.

Other modifications of Captain Noble's gauges were soon introduced for the experiments with submarine explosions made in foreign countries. Thus in France it is understood that a gauge made somewhat as shown on Fig. 7 was used in a number of experiments, the details of which, however, have been carefully kept secret. The gauge consisted of a hat-shaped metal body *a*, *b*, *c*, *d*, *e*, with an eye-bolt *f*. Rubber washers *g g*, a steel piston *k*, a rubber diaphragm *l*, a ring washer *m*, and clamps *n n*. The lead pellet *h* of dimensions figured was much larger than those used in England.

It was not very difficult to improve upon this gauge, and Captain Eckermann, of the Swedish Engineers, did this by adopting the form of piston used in one of the English gauges already described, using an india-rubber washer instead of a diaphragm, decreasing the diameter of the ring washer, and lipping it so as to engage the rubber between it and the top of the piston, and using a screw cap in place of the clamps and steel centering pins instead of the two rubber rings round the pellet. He thus produced the gauge shown in Fig. 8; three gauges were placed back to back at 120 deg., the whole forming the crusher gauge now known as "Eckermann's." It costs 4*l.*, and 1000 lead pellets with necessary tools cost 16*l.*—total for six and 1000 pellets, 40*l.* The great defect in this crusher gauge is the same as that in all the European forms of Noble's gauge, viz., that no provision is made to prevent the piston jumping in and out, and hammering the pellet by a series of blows which are not always given in the direction required. Pellets are sometimes extracted from these gauges, as shown in sketch, Fig. 9, indicating that after the first blow the piston has *jumped back* and released the pellet from the centering pins; the pellet has then toppled and received a second blow while in a tilted position, and then several smaller blows. The centering pins are evidently inferior to the rubber rings used in the English and French gauges for the same purpose, and the necessity of some arrangement, as in the American gauges, for preventing the piston from moving backwards is very apparent.

The records from these gauges, when applied to submarine explosions, are distinctly inferior to those obtained in the English experiments with small pellets of copper or lead. Thus, taking haphazard one of the English experiments, a 13-in. shell fitted with four copper pellet gauges gave .019, .015, .021, .018, and four lead pellet gauges in the same shell gave .166, .174, .166, .170. In some records with Eckermann's gauge we find in one triple gauge .055, .065, .025, differing more than 100 per cent. In another experiment a triple gauge gave .095, .103, .079. In another, .105, .105, .087. In another, .091, .130, .115.

Very few experiments with the piston and cylinder type of crusher gauge have been made in England since the conclusion of the experiments against the Oberon. But a tubular form of dynamometer has been adopted into the service which in its earliest form, shown on sketch (see Figs. 10, 11, and 12), was invented and proposed by the author in February, 1876. It acts on the collapsible principle, and the work performed upon it is measured by the difference in the amount of its cubical content before and after an experiment.

The results obtained with the first form of this dynamometer were encouraging; and it has gradually been improved by the inventor until it has now become trustworthy. In the latest (1886) patterns the tubes are made of commercially pure lead obtained by the desilverising process. The tubing is run from a special die of elliptical section (Fig. 13) $1\frac{3}{4}$-in. major axis and $1\frac{5}{8}$-in. minor axis, with a circular core

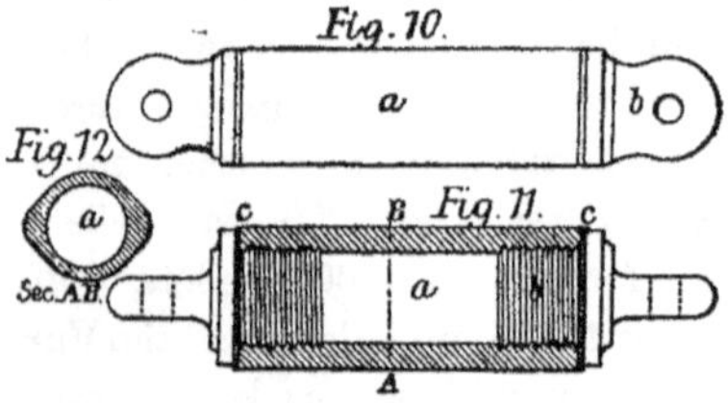

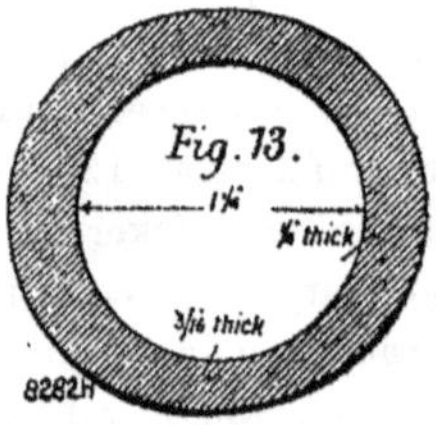

$1\frac{1}{4}$ in. in diameter. The tubing is cut into short lengths, straightened and trimmed to exactly 6 in. in length with a smooth, flat, and square surface at each end. They are then packed in perforated wooden slabs in boxes so that they do not touch, and cannot damage each other or be damaged in transit; fifty tubes are placed in one box.

When used to measure the effects of explosions, three tubes are placed in a cage formed as follows. The top consists of a circular plate of $\frac{1}{2}$-in. iron $4\frac{1}{2}$ in. in diameter with a central $\frac{11}{16}$-in. hole. To this is secured at one end an iron cage 8 B.W.G. thick, 7 in. long, and perforated with seven rows of $\frac{7}{8}$-in. holes, seven holes per row. The bottom plate, loose, is similar to, but $\frac{1}{8}$ in. smaller in diameter than, the top plate. (See Figs. 14 and 15.)

For each plate is provided an india-rubber disc $\frac{1}{8}$ in. thick, $4\frac{3}{8}$ in. in diameter, with a central $\frac{3}{4}$-in. hole. A $\frac{5}{8}$-in. eye-bolt with a nut at the lower end secures the whole together, three tubes being previously inserted between and square to the india-rubber discs. Care should be taken not to screw up one cage of tubes more tightly than another. These cages are usually suspended by a small wire rope, with eye and shackle, from floating spars, the distance from the charge being measured on the spar, and the required submersion on the wire rope. The cages cost 5*l.* per dozen, and the tubes cost 19*l.* per 1000, but the old lead after the experiments can be sold for about 12*l.* per 1000. Hence the cost of the tubes does not exceed 7*l.* per 1000. The pellets for Eckermann's gauges are much lighter. Comparing cost:

12 of the above cages and 1000 tubes	12*l.*
12 Eckermann's holders and 1000 pellets	54*l.*

FIG. 14

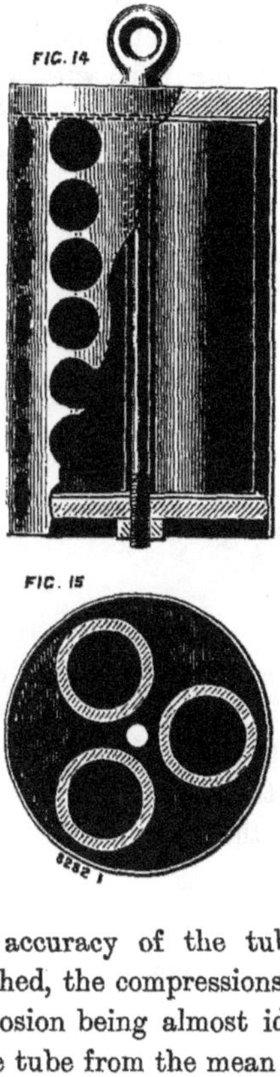

FIG. 15

Fig. 16.

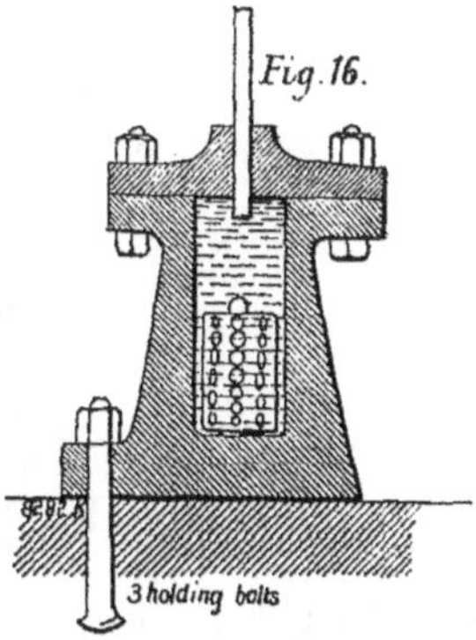

The accuracy of the tubular dynamometers has been thoroughly established, the compressions of the three tubes in a cage subjected to an explosion being almost identical. Thus the greatest difference of any one tube from the mean of three in the cage was found to be in six cages, and in one of the large charge experiments as follows: Greatest difference from mean, cage (*a*), 1.6 per cent.; (*b*), 3.2 per cent.; (*c*), 0.2 per cent.; (*d*), 2.4 per cent.; (*e*), 4 per cent.; (*f*), 2 per cent.

The best way to measure the tubes is to get some carefully washed sand not too fine (silver sand is too fine), and dry it and let it cool. Place a tube on a small piece of cardboard, quarter fill with sand and tap the tube, half fill and repeat tapping, three-quarter fill ditto, fill and tap, and fill and tap until no more sand can go in. Then strike top surface level, and weigh the sand carefully = W. After the experiment repeat = W^1. Then $W - W^1$ = compression. And $x : 100 :: W - W^1 : W$. x is the percentage of capacity compressed, and forms the method of comparison adopted; it is independent of the specific gravity of the material used for measuring the compression.

In order, if possible, to dovetail the records from these dynamometers with those from the piston and cylinder type of crusher gauges, experiments have been carried out to find the compressions produced when the tubes were placed in a closed cylinder filled with water and subjected to the hydro-dynamic effect produced by a weight falling upon a small movable plunger. A very strong metal cylinder 5 in. in dia-

FIG. 17

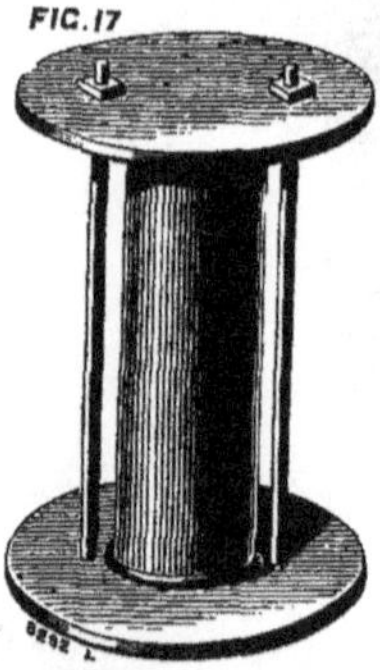

meter and about 1 ft. long internal dimension was used, and the cylinder cover was fitted with a movable plunging rod having a sectional area of 1 square inch. (See Fig. 16.)

The fall of 10 lb. through 10 ft. on this plunger produced the following curious records on the tubes when three of them in one cage were subjected to the blows:

				Mean.
Trial 1.	$W - W^1$ = 479	61	32	191
,, 2.	,, 39	228	485	217

These unexpected results proved how very different is the blow given by an explosion to that given in the experiment. In the one the three tubes are compressed nearly equally, in the other the compression of the weak tubes saves the others.

It occurred to the author that the trial should be made with one tube only in the cylinder, and a special cage was made (for a single tube) consisting of two plates and rubber discs braced together by two bolts, one on each side of the tube to be tested (see Fig. 17). A blow of 100 foot-pounds on the plunger, caused by the fall of a 20-lb. weight, now gave in three separate trials on three tubes taken singly the following values: $W - W^1 = 230$, 314, 288, mean 277; or, 14.66 per cent. of $W = 1900$. Continuing the results of trials with 20-lb. falling weight:

TABLE VI.

	ft.		ft.-lb.			Mean Value.
Falling	2	=	40	gave x, or per cent. of W =	2.9, 2.37, 3.0	=2.76
,,	3	=	60	... ...	7.37, 7.26, 7.68	=7.44
,,	4	=	80	... ...	9.42, 11.47, 11.21	=10.7
,,	5	=	100	... ...	12.21, 16.62, 15.16	=14.66
,,	6	=	120	... ...	21.26, 20.21, 20.42	=20.63
,,	7	=	140	... ...	23.0, 25.42, 24.58	=24.33
,,	8	=	160	... ...	30.26, 30.68, 32.37	=31.10
,,	9	=	180	... ...	40.0, 34.68, 38.37	=37.68
,,	10	=	200	... ...	40.0, 43.05, 43.68	=42.26

The compressibility of water was clearly shown by the monkey being thrown back to a height of from 15 to 20 per cent. of the fall. It was then caught and not allowed to fall a second time on the plunger. The above values were slightly increased by halving the quantity of water in the cylinder, by the introduction of a large lump of iron about 5 in. in diameter and 6 in. long. The result was evidently due to less energy being absorbed in compressing the water, there being less water to press. It would be very interesting to find by experiments whether the above values can be equated with hydrostatic pressures applied to a similar or to the same cylinder, the hole for the plunger being closed by a screw plug. The values just given when plotted graphically on a diagram where the abscissæ are foot-pounds applied, and the ordinates the values for x obtained, produce a line almost straight.

The lead pellets for the Eckermann's crusher gauges have been subjected to similar blows given by a weight of 22 lb. 1 oz. falling through various heights. Each pellet is 55 mm. long and 20 mm. in diameter, or 0.4856 square inch sectional area.

TABLE VII.—COMPRESSIONS—LARGE LEAD PELLETS.

Foot-Pounds.	Compression in Inches.	Foot-Pounds.	Compression in Inches.	Foot-Pounds.	Compression in Inches.	Foot-Pounds.	Compression in Inches.
20	.13	120	.51	220	.705	320	.84
40	.23	140	.56	240	.73	340	.87
60	.31	160	.60	260	.755	360	.89
80	.38	180	.64	280	.78	380	.93
100	.45	200	.68	300	.81	400	.96

The pellets used in the English crusher gauges were also subjected to similar blows from a falling weight of 25 lb. In the following Table the blow in foot-pounds and the compressions of a copper pellet $\frac{1}{2}$ in long and $\frac{1}{12}$ square inch section are given, also a column of means, and the same corrected by a curve. Also in the last column are recorded the actual statical pressures which by other experiments were found to produce the same compressions.

TABLE VIII.

Twenty-five Pounds Fall in Inches.	Foot-Pounds.	Compression in Inches.					Pressure Producing the same Compression.
		First.	Second.	Third.	Mean.	Corrected.	lb.
3		.036	.038	.043	.039	.039	2365
6	12.5	.058	.059	.063	.060	.060	3050
9		.066	.070	.079	.072	.078	3615
12	25	.090	.092	.094	.092	.094	4095
15		.104	.105	.105	.105	.108	4515
18	37.5	.116	.116	.117	.116	.121	4906
21		.125	.128	.136	.130	.133	5305
24	50	.143	.145	.148	.146	.144	5645
27		.148	.151	.152	.151	.154	5965
30	62.5	.160	.162	.166	.163	.163	6280
33		.167	.168	.169	.168	.171	6535
36	75	.177	.178	.179	.178	.179	6795
39		.184	.188	.189	.187	.187	7070
42	87.5	.190	.191	.191	.191	.194	7295
45		.200	.201	.203	.202	.201	7540
48	100	.204	.207	.207	.206	.208	7820
51		.208	.211	.214	.211	.215	8100
54	112.5	.218	.221	.223	.221	.222	8380
57		.223	.226	.233	.228	.228	8620
60	125	.231	.232	.237	.234	.234	8860

The numerous crusher gauge experiments made by the War Department have never been published, nor is the author permitted to do so. Many of them were conflicting and difficult to explain, many on the other hand were interesting and suggestive.

CHAPTER III.

Theoretic and Empiric Formulæ for Submarine Explosions.

Early in 1874 the author pointed out that the results of certain crusher gauge experiments in England appeared to indicate that the effects of submarine explosions, as shown on the gauges, varied inversely as the cube of the distance between the centre of the charge and the surface of the target. Lieutenant (now Major) English, R.E., then examined this theory, and wrote the following very interesting and important remarks thereon, dated March 23, 1874:

"It appears that a permanent compression of 0.234 in. is produced by the blow given by a weight of 25 lb. falling through 60 in. on a copper cylinder 0.5 in. long and 0.083 of a square inch in area. Also that an equal compression is produced by a steady pressure of 3.95 tons upon a similar cylinder.

"Assuming that the maximum pressure produced during the impact of a falling weight varies with the square root of the height through which the weight falls,* and that, as above, the maximum pressure produced by a weight of 25 lb. falling through 60 in. on to a copper cylinder of the dimensions given is 3.95 tons, the results given in the following Table are obtained:

Table IX.

Height of Fall in Inches.	Calculated Maximum Pressure in Tons.	Calculated Compression in Inches.	Observed Compression in Inches.
60	3.95	0.234	0.234
54	3.79	0.224	0.222
48	3.57	0.212	0.208
42	3.35	0.199	0.194
36	3.10	0.184	0.179
30	2.83	0.166	0.163
24	2.57	0.148	0.144
18	2.19	0.120	0.121
12	1.80	0.091	0.094
6	1.29	0.053	0.060

"Within the limits of the experiments it may, therefore, I think, be

* *Vide* Royal Engineer Corps Paper X., vol. xviii., published 1870, and dated September 27, 1869, by Lieutenant T. English, R.E., "On the Statical Pressure produced by the Impact of a Falling Weight."

assumed that the greatest pressure produced by a weight falling upon similar cylinders varies as the square root of the height from which it falls, that is, directly as its striking velocity.

"In the explosion of a torpedo, assuming the variations of pressure to be transmitted outwards at the same velocity in all directions from the charge, it is clear that the velocity of any particle considered to lie on the surface of a sphere, of which the charge is the centre, will vary as the $\frac{\text{surface}}{\text{volume}}$ of the sphere, that is, inversely as the radial distance from the charge."

This sentence has recently been more fully explained to the author by Major English as follows :

Consider a thin spherical shell with centre at C C being the charge. Let A B D E be any portion of it bounded by lines radiating from the charge. Then, assuming water to be incompressible, and the maximum

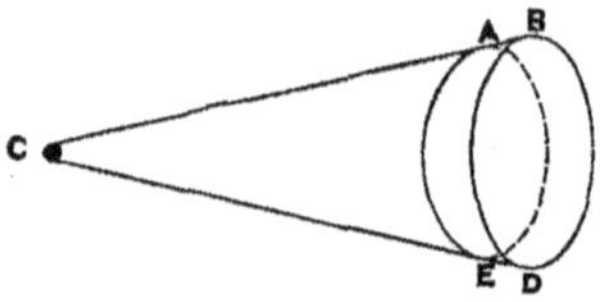

hydrostatic pressure due to the explosion to be p pounds per square inch, the pressure on A E will be $p \times$ surface A E, tending to cause motion outwards. The mass of water set in motion will be that of the volume B C D. By the equation $P = mf$,

$$p \times \text{surface A E} = \text{vol. B C D} \times f,$$

consequently f varies as $\frac{\text{surface}}{\text{volume}}$, which is proportional to that of a sphere of which the charge is the centre.

Also as the time through which p acts is the same at any distance from c,

$$v = f t, \text{ and also varies as } \frac{\text{surface}}{\text{volume}}.$$

In the above, f = acceleration, or increase of velocity in one second.

"If the effect on copper cylinders be considered to be produced by the blow of an uniform weight of water striking them, and if the maximum pressure produced follows the law already shown to hold good for a falling weight, and varies directly as the striking velocity, it follows that the maximum pressure per square inch produced by the explosion of a torpedo is, for bodies of equal resisting power, inversely as their radial distances from the charge.

"From the experiments (October and December, 1873) with 500 lb, charges of gun-cotton it appears that the results upon similar cylinders at various distances were as follows:

TABLE X.

Letter Indicating Cylinder.		Radial Distance from Charge in Feet.	Observed Compression in Inches.	Calculated Pressure in Tons per Square Inch from Table.	Pressure Multiplied by Radial Distance.
In same radial line . . .	B	23	0.055	7.05	162
	C	30	0.024 (bad)	4.50	135
	D	37	0.020	4.10	152
	E	44	0.014	3.50	154
Ditto . . .	F	30	0.066	8.20	246
	J	60	0.018	3.80	228
	G	38	0.047	6.60	251
	H	45	0.032	5.40	243

"The results observed from F J G H show a tolerable accordance with the law given above, and although cylinders B C D E were arranged along the bottom, and therefore under different conditions to the others, yet they agree fairly among themselves.

"Within the limits of the experiments, the curve of which the compressions of the copper cylinders are abscissæ, and the corresponding pressures ordinates, approximates to a parabola with its axis on the line of abscissæ. Hence, for this part of the curve, the area which represents the work done on the cylinders varies as the cube of the pressure, that is, inversely as the cube of the radial distance from the charge.

"Beyond these limits, however, and under greater pressures, the curve diverges altogether from the parabolic form, and this law would no longer hold good."

The theory explained and formulated above, viz., that the pressure produced at any distance from an explosion varies directly as the distance, has been adopted by the author in a formula to be discussed at the end of this chapter, a term being added to express the high pressures met with at small distances from the charge. As stated on page 15, the crusher gauges used in England were screwed into iron shot or shell suspended by floats and attached to the bottom by sinkers; or they were screwed into rings which were fixed to sinkers, and therefore rested on the bottom; or they were secured to an iron or metal ring surrounding a small charge placed in the centre, as in the American experiments; or they were secured to the side of a vessel subjected to the blow of a submarine mine or torpedo. The gauges in the shells pointed in all directions, and it was noted that when such a shell was placed at a fair distance from the exploding charge, the compressions in all the gauges were not unfrequently alike, and seldom varied greatly *inter se*.

In the ring experiments both in England and America the compressions obtained in the top crusher gauges were usually greater than the compressions obtained on the gauges in the same horizontal plane as the charge, and still greater than the compressions in the bottom gauges of the ring.

This difference was moreover more observable when the ring was near the surface than when it was submerged to a greater depth, indicating, apparently, that the greatest pressures were produced towards the lines of least resistance.

At very small distances the results are abnormally severe. This was observed in many of the American experiments with small charges in rings fitted with gauges, and was especially observable when the higher explosives were employed.

As before stated, the large pellets in some experiments appear to have been compressed by more than one blow, frequently by two severe blows of almost equal intensity, and then by a series of smaller blows. The second blow may be due to the reaction from the bottom, but it is not easy to account for the smaller blows following it.

When the explosion of a charge takes place a sphere of gas is formed at an enormous pressure, which probably varies directly as the charge and as the intensity of action of the explosive. This sphere of gas expands quickly or slowly according to the amount of resistance opposed to it, as shown by the greater height to which the water is thrown by charges exploded near the surface, a smaller mass of water being driven at a greater height, and consequently at a greater initial velocity.

Moreover, the crater of upheaval increases in diameter with the submersion of the charge up to a certain limit, the work being expended in moving a larger mass of water at a smaller velocity. The sphere of gas, whatever be the position of a charge, must expand more rapidly in the direction of the line of least resistance (L.L.R.), and in so expanding pushes before it, with immense force, the mass of water lying in and around the L.L.R. If a vessel happen to lie in or near to this L.L.R. she must receive the shock due to the *vis viva* of the whole or part of this water in motion, and she succumbs to this racking blow, or resists it according to her strength. If the L.L.R. happen to pass through the ship's side, *i.e.*, if her strength offer less resistance to the force of the explosion than the inertia of the water between that part of the ship's side and the surface, then must her side or bottom be of necessity blown in, and this consideration alone shows how much more effective an explosion must be when directly under the keel of a vessel than when under her side. Yet the initial pressure sustained by the nearest portions of the vessel to the charge is the same in these two cases, although the well-known results are widely different.

If the history of submarine explosions, whether in war or peace, be examined to present date, no single instance will be found in which an iron vessel possessing a skin at all approaching in strength that of the outer skin of a modern warship has sustained any fatal injury from an explosion when she lay outside the water crater. The vessel may receive a shock that discovers weak points in her machinery or hull, but if sound, and properly built, she will apparently receive no fatal injury unless subjected to the blow of water in mass driven against her at high velocity.

It is extremely probable that the pressures registered in crusher gauges, placed in different positions with reference to a charge, may assist in solving the problem as to where a vessel of given strength is likely to receive a fatal injury from such charges. They enable us to measure the pressures, but we cannot measure the dimensions of the globes of gas suddenly formed (except very roughly by the craters made in mud), nor can we directly measure the velocity at which the water is driven against a ship's side or bottom. Moreover, inasmuch as it is impossible to try each kind of explosive against a target representing an ironclad, crusher gauge experiments became convenient methods of comparing the comparative intensity of action of different explosives. But crusher gauge experiments must be received with great caution, as well as any formulæ deduced from them. The mathematical exactitude which seems to be allied with a formula may do more harm than good unless it be verified by target experiments.

A collection of facts, as recorded from actual target experiments, must be the first thing to examine carefully, and the tabulated results already given are most useful in this direction. In short, when the limits of submarine mining experiments on which an empiric formula is based are not exceeded, it is more likely to be trustworthy than one which is founded on theoretical considerations, because our knowledge of what occurs during a subaqueous explosion is at present so restricted.

The experiments carried out in America were mostly with small charges, and are chiefly useful in giving an indication as to the relative intensity of action of various explosives, as registered by crusher gauges placed at small distances from the charges.

The careful and scientific investigation of the theory of subaqueous explosions which General Abbot has made, rests mainly on such experiments, and his deductions must therefore be accepted with caution. They are not always corroborated by the results of experiments with larger charges, as was shown in some of the explosions fired against the Oberon, and against the target ship at the Carlskrona experiments.

Every expert should study General Abbot's book, "Submarine Mines, 1881," with the greatest attention, but it is probably out of the reach of many who will read these papers, and a summary of his investigations and formulæ will therefore be given.

The most important deductions from the numerous experiments carried out by General Abbot in America are as follows:

Let

W = the mechanical work done by a submarine explosion, expressed in foot-pounds per square inch of surface exposed to the shock.

P = intensity of action, or sudden pressure in pounds per square inch of above surface exposed.

C = weight of charge in pounds.

D = distance in feet from centre of charge to above surface.

S = submersion of centre of charge in feet.

N = number of fuzes in the charge, distributed uniformly.

R = radius of a sphere equal in volume to the explosive fired by each fuze.

N.B.—With two fuzes attached to 12 ft. of lightning fuze coiled in a mine charge, experiment proved that $R = \frac{C}{28}$.

a = angle in degrees at centre charge between the vertical downwards and the line of direction of the blow examined.

E = value to be found by trial for each explosive compound.

M = ditto for each explosive mixture.

Then for explosive compounds:

$$W = \frac{0.21(a+E)C}{(D+0.01)^{2.1}} \quad . \quad . \quad . \quad (1)$$

and

$$P = \sqrt[3]{\left(\frac{6636\,(a+E)\,C}{(D+0.01)^{2.1}}\right)^2} \quad . \quad . \quad . \quad (2)$$

Also for explosive mixtures:

$$P = \sqrt[3]{\left(\frac{MS^{\frac{1}{40}}\,C^{1.94}}{(D+1)^2\,\sqrt{R}}\right)^2} \quad . \quad . \quad . \quad (3)$$

The following considerations were also evolved. Remarks on them by the present author are bracketted as being more convenient than a number of foot-notes.

Other variables may be mentioned, such as the depth of water under the charge, the character of the bottom, &c.; but as the experiments at Willet's Point, where the bottom was soft mud, appeared to indicate that these had no sensible influence on the results, they are neglected.

[This is not borne out by experiments elsewhere. For instance, in the Oberon experiments with 500 lb. charges of gun-cotton it was considered that the effects of No. 5 experiment were no greater than those of No. 4, and that the effects of No. 6 were decidedly greater than those of No. 5. It is considered by some experts that large ground mines are more effective than equal charges buoyed up considerably from the bottom by nearly 30 per cent., but this is very doubtful.] Again, the nature of the priming might be expected to influence the force produced by the explosion of an explosive mixture; "but the results with gunpowder appear to be sensibly the same whether the priming consists of fulminating mercury, safety compound, or gunpowder." [In an addendum General Abbot draws attention to the announcement made by Messrs. Roux and Sarran so long ago as 1874, that they had succeeded in obtaining from musket powder more than four times the usual explosive force by the use of primers so large as to apply a detonating shock to all parts of the charge. This has not been verified by experiment.] In framing a general formula for submarine explosions it may fairly be assumed that the available energy developed by the detonation of higher explosives varies directly as C. With mixtures, however, the case is different, because with a weak envelope some of the powder is driven into the water unexploded, and with a strong envelope, although this loss is minimised, energy is consumed in the rupture thereof. With mixtures therefore the energy may be said to vary as C^x. "In a perfectly incompressible fluid, the total energy transmitted through it from molecule to molecule must be equal upon every spherical surface enveloping the explosive as a centre. In other words, the energy at any point must be inversely proportional to the square of the distance of that point from the centre of the charge. Water, however, is not a perfectly incompressible fluid. Moreover, a formula framed upon the supposition that the energy varies only as a power of D would indicate an infinite energy for zero distance. For these reasons the function for distance should take the form $(D + A)^q$ in which A is a small constant, and in which q must always be nearly equal to 2. This function of course enters the formula for W in the denominator. [It would appear from Major English's remarks, already given, that the equation connecting the distance with the mechanical work done cannot be expressed in so simple a manner for all distances; and if the limits be confined to those which are of practical value, that the exponent is more nearly 3 than 2.] The effect of increasing S is to increase the fluid pressure round C, and therefore to increase the resistance to the formation of the globe of gas when C is exploded. A function S^y in the numerator satisfies these conditions. Again, when $R=0$ $C=0$ $\therefore$ R^z placed in the denominator will fulfil the needful conditions for that quantity.

"The probable action of the forces developed by a subaqueous explosion indicate that the normal line of maximum intensity will be directed upward, and of minimum intensity downward. A function of the form $(a+E)^t$ in the numerator will fulfil these conditions."

[If this should satisfy the conditions of various submarine explosions, as regards E E^{I}, &c., for different explosives, it evidently gives an extremely high relative value to the higher explosives, *i.e.*, those in which E is large as compared with a, and as compared with E^{I} E^{II}, &c., for other and lower explosives. The question arises do they possess this high relative value, when for instance employed in large charges acting at a distance? They do possess it when examined for energy developed at small distances from the charge, and the values for E E^{I}, &c., have been found by an examination of these actions, *i.e.*, by crusher gauge records on 3 ft., 4 ft., and 5 ft. rings surrounding small charges. An examination of the American experiments shows that the crusher gauge experiments at increased distances were apparently only carried out with one explosive, dynamite. In these the gauges were fixed to an iron crate or framework, so that the most distant gauges were 25½ ft. from C, and the latter was gradually increased up to 100 lb. dynamite, the pressure per square inch thereby produced being recorded as 3088 lb. Also there were a few experiments with large charges, in which crusher gauges were used at greater distances; but they were placed on the bottom, and the blows reflected. Experiments in England proved that the effects in crusher gauges so situated were considerably less than on similar gauges fixed, even to movable objects, but buoyed up from the bottom. A formula therefore which satisfied the pressures recorded in the former would fail entirely to account for the much higher pressures recorded in the latter.]

General Abbot's formula for the most general case of a subaqueous explosion then takes the form :

$$W=\frac{K(a+E)^t\ S^y\ C^x}{(D+A)^q\ R^z}$$

K being a constant varying with the nature of the explosive.

Lengthy calculations based principally on a number of crusher gauge experiments with small charges evolved from this general equation the three special equations already given and numbered (1), (2), (3).

By equation (1) P varies directly as the two-thirds power of C, and inversely as the $\frac{4.2}{3}=1.4$ power of D, whatever may be the direction of the object or target. Also P varies directly as the two-thirds power of E when the target is over the charge, α then being 0.

By equation (2) W varies directly as C and inversely as D^2 nearly.

Equation (3) refers only to the horizontal plane through the centre of the charge. Hence if we wish to compare the compounds with the mixtures, α in equations (1) and (2) must $= 90$ deg. The values of M for different explosive mixtures, found after a large number of experiments in America, are:

Table XI.

Mortar powder	M = 790
Mammoth powder	M = 20
Cannon powder	M = 59
Musket powder	M = 1,554
Sporting powder (fine orange lightning)	M = 3,331
Safety compound, Oriental Powder Company, a mixture of potassium chlorate and gambier ...	M = 13,383

Applying equation (3), let us find P when C = 100 lb. safety compound, and when S = 5 ft., and D = 10 ft. Then as M = 13,383,

$$\text{Log } P = \tfrac{2}{3}\left(\log 13{,}383 + \tfrac{1}{40}\log 5 + 1.94\log 100 - 2\log 11 - \tfrac{1}{2}\log 0.8\right)$$

and

$$P = 9842 \text{ lb. on square inch.}$$

Safety compound, although so powerful as compared with the other explosive mixtures, "has been driven from the market by the nitro-compounds which have been proved to be both stronger and safer to handle."

[It may, however, sometimes occur that officers are compelled to use gunpowder. Its effects at different distances are consequently calculated further on.]

Before leaving this part of the subject General Abbot noted that:

1. The strength of gunpowder for submarine work is nearly inversely proportional to the size of the grains.

2. Large charges of gunpowder are less wasteful than small charges, because a smaller proportion of the charge is blown into the water unburnt. [This depends greatly on the mode of ignition.]

3. A strong case is required for a charge of gunpowder for a similar reason.

4. No advantage is obtained by the employment of detonating fuzes or of a large detonating primer. [Questionable.]

5. An air chamber in a gunpowder mine placed between the charge and the object is highly advantageous, because it directs the blow and increases the effective pressure in the desired direction.

6. An excellent way to ignite a large charge of gunpowder is to use two fuzes in connection with some lightning fuze coiled away in the powder.

Detonating Compounds and Mixtures.

Before examining and comparing the different explosives which have been tried in the American experiments it will be convenient to record some of the more important general conclusions that were discovered during the American series of experiments.

1. *Submersion.*—So long as a charge is submerged to an extent likely to occur in practice, the results on a target are not affected to any appreciable extent. In other words, S has so small a fractional exponent as to become practically unity, and may therefore be omitted as a multiplier in the formula. A submersion of 4 ft. for a charge of 100 lb., and 4 ft. per 100 lb. for larger charges, being sufficient to develop the full effects of a subaqueous explosive of the first order.

2. A thin weak envelope gives the best results. [Just the reverse with gunpowder, as already noted.]

3. No advantage is gained by the use of more than one detonator. [Again different to gunpowder.]

4. The energy cannot be directed by the employment of air chambers in the mine. [Again different to gunpowder.]

5. An air space not exceeding three times the volume of the charge complete with its apparatus, &c., has no prejudicial effect. [A much larger air space is required in practice.]

6. The mine case should not be formed of compressible material that can absorb work by pulverisation, such as wood [which differs from air in not again giving forth the energy absorbed in its compression].

Two-inch wooden cases for small charges of dynamite occasioned a loss of effect on gauges of no less than 55 per cent.; with dualin the loss was 47 per cent., and with gun-cotton 40 per cent.

7. Some explosive compounds are liable to sympathetic explosion. Dry is much more sympathetic than wet gun-cotton. Compacted dynamite is more sympathetic than loose dynamite. Abbot's equation for finding the distance of sympathetic explosions under water is:

$$D = B^{2.1}\sqrt{C}$$

where B is a value found by experiment with each explosive and description of envelope, and D is the distance in feet from a charge C lb. at which sympathetic explosion will occur. With compacted dynamite and thin tin envelope $B = 20$.

8. "The considerable effect upon the numerical value of P produced by varying the direction of the line of action in vertical planes passing through the charge and vessel, emphasises the importance of placing the former under the latter if possible.

9. "The small exponent of C (only $\frac{2}{3}$) in the value of P shows that with a given weight of explosive many moderate charges block a channel more effectually than a few large ones. Thus a mine containing 500 lb. is only 2.9 times as effective as one of 100 lb." [The exponent of C in the formula, and the theory resting upon it, are equally open to doubt.]

10. Explosives which are compacted are more difficult to detonate with a fuze than those which are in a looser state. Thus loose dynamite, even if in the frozen state, can be always detonated by 15 grains of mercurial fulminate in a copper capsule, whereas compacted dynamite in the frozen state cannot. It is due to the larger surface exposed to the flame and shock of the fuze when the dynamite is loose. [It is to be observed that the compacted dynamite is the most sensitive to sympathetic detonation, and although this at first sight seems contradictory, it may be accounted for similarly, a larger amount or surface of dynamite being then in contact with the envelope through which the detonation is transmitted.]

11. Some of the compounds are equally powerful when wet and fully exposed to the sea water as when dry. [General Abbot claims that this fact was discovered by him before wet gun-cotton was first detonated in England, late in the year 1872.]

12. Some of the compounds are both safe and effective when in the frozen state, and are not deteriorated when subjected to great differences of temperature in store.

[The coefficients for the intensity of action for the different explosives as stated may probably be accepted as giving fairly accurate results when applied to General Abbot's formulæ (1) and (2) within the limits of the experiments on which they are based. But the pressures calculated for longer distances by equation (2) do not agree with the results of the experiments made in this country. For instance, the 500 lb. experiments in October and December, 1873, already tabulated in the quotation from Major English's memorandum, show that the pressures vary within the limits of those experiments inversely as the distances, and not inversely as the 1.4 power of the distances as suggested by General Abbot.]

The formula proposed by the writer will be examined in a subsequent page.

It is, of course, important that correct views should be held concerning the manner in which submarine explosions act, in order that the formulæ required for determining the various effects produced by different explosives and charges at different distances, and under different conditions, may be approximately determined.

An examination of experiments will often be sufficient to guide practice without any formulæ; but as new explosives come on the field, and fresh conditions become involved, a correct formula becomes more than ever necessary if the mines are to be arranged and charged in a good and economical manner. Evidently the initial pressure must vary as the quantity of gas suddenly generated at a high temperature.

It would be extremely difficult to estimate the quantity of permanent gas suddenly produced and the temperature of explosion, and still more difficult to introduce a function for increased effect under water due to the greater or less sudden action of any given explosive as compared to another explosive; but experiments have given some fairly accurate data on the relative intensity of action under water of different explosives, and in all probability this intensity of action includes in the most practical form both the gas suddenly generated and its temperature, as well as the relative velocity of detonation, and perhaps other important but complicated matters which surround the subject of subaqueous explosions.

Assuming that the maximum pressure produced by an explosion varies directly as the intensity of action of the explosive as compared with other explosives in experiments made with small charges acting under water at small distances, it is evident that it also varies as the quantity of explosive used in a charge, or as $C \times I$.

Adopting the ruling idea set forth in Major English's memorandum already quoted, viz., that between certain limits the pressure produced by the explosion of a torpedo varies inversely as the distance of the object, we arrive at the conclusion that the pressures vary approximately as $\frac{C \times I}{D} \times$ constant.

C being the weight in pounds of explosive in charge.

I being the relative intensity of action of the explosive: dynamite No. 1 being the unit.

D being the distance in feet from the centre of the charge to an unscreened crusher gauge.

The pressure per square inch, as recorded in gauges, being found for various distances, and charges of explosives, an examination proved that at moderate distances the equation $P = \frac{C\,I}{D} \times$ constant was fairly well satisfied when constant $= 9$.

But this equation does not agree with practice at small distances.

Close to the charge, say within 20 ft., different conditions obtain, and it therefore becomes necessary to add to the above simple form of equation for pressure a term which would satisfy approximately the greatly increased pressures recorded at the smaller distances.

It appeared probable that this term must also vary directly as $C \times I$, and that it should take the form $\frac{C\ I}{D^x} \times$ constant. After much labour the writer found that fairly accurate results are obtained in the horizontal plane when $x = 3$, and the constant $= 225$.

The equation consequently becomes:

$$P = \frac{9\ C\ I}{D} + \frac{225\ C\ I}{D^3}$$

or

$$P = \frac{9\ C\ I}{D}\left(1 + \frac{25}{D^2}\right) \quad . \quad . \quad . \quad . \quad . \quad (4)$$

Applying this equation to some of the American experiments with dynamite charges, and to some of the English experiments with gun-cotton charges, it will be seen that it gives pressures agreeing more closely with the pressures recorded on the crusher gauges than do the pressures computed by General Abbot's formula.

TABLE XII.

C.	I.	D.	P Pounds per Square Inch.			REMARKS.
			Observed on Gauges.	Calculated.		
				Abbot's Formula.	Formula now Proposed.	
lb.		ft.				
1	100	1.62	8000	7,555	7853	Dynamite.
10	100	3.62	8500	11,432	7458	,,
50	100	25.5	1840	2,181	1835	,,
100	100	70	1141	842	1293	,,
100	100	25.5	3088	3,461	3671	,,
200	100	80	1365	1,109	2260	Crushers on a ground mine. Blow apparently deflected by the bottom.

The American experiments proved that the mean pressures in the vertical plane were greatest towards the zenith and least towards the nadir, also that these differences varied inversely (nearly) as the intensity of action of the explosive. The pressures computed by the equation

$$P = \frac{C\ I}{D}\left(1 + \frac{25}{D^2}\right)$$

are for the horizontal plane. If the target be above or below this plane the pressures should be increased or decreased by the following percentages (e) according to the explosive employed:

For gun-cotton, dynamite No. 1, and other explosives whose intensity of action is about 100, add or deduct 20 per cent. at the vertical, and proportionately smaller amounts for other angles smaller than 90 deg. from the horizontal plane.

For blasting gelatine, whose intensity of action is about 142, add or deduct 12 per cent. at the vertical, and smaller amounts proportional to smaller angles. And so with other explosives, as shown in the following Table for I and *e*. For gunpowder the writer takes $I = 25$ and $e = 35$.

The above assumes that the charges are sufficiently tamped or submerged, say by at least 4 ft. per 100 lb. of explosive in the charge. When submerged to a smaller extent the pressures, especially for gunpowder charges, cannot be computed, so much of the force being expended in driving water uselessly into the air. For a similar reason the computed pressures form no indication of the results when a charge is placed close to a weak vessel—in this case, however, it is most usefully employed.

If, therefore, P be the pressure in pounds per square inch on a target in the vicinity of the submarine explosion of a charge containing C lb. of any explosive whose intensity of action under water is I; then if the target and charge be in the same horizontal plane at distance D ft. apart,

$$P = \frac{9\,C\,I}{D}\left(1 + \frac{25}{D^2}\right) \quad . \quad . \quad . \quad . \quad . \quad . \quad (4)$$

and if the target be out of the horizontal plane,

$$P = \frac{9\,C\,I}{D}\left(1 + \frac{25}{D^2}\right)\left(1 + \frac{\beta}{90} \times \frac{e}{100}\right) \quad . \quad . \quad . \quad . \quad (5)$$

β being the angle between the line joining the centre of the charge and target and the horizontal plane, and *e* being the percentage for the particular explosive used, plus if target be above horizontal plane through charge, minus if below. The values taken for I and for *e* are:

TABLE XIII.

Description of Explosive.	I.	*e*.
Blasting gelatine ...	142	12
Forcite ,, ...	133	14
Gelatine dynamite No. 1	123	16
Dynamite No. 1... ...	100	20
Gun-cotton	100	20
Gunpowder	25	35

Applying this equation (5) to the 500 lb. gun-cotton experiments already quoted in Major English's memorandum, and comparing the results with the pressures calculated by General Abbot's formula, the following pressures are obtained, again pointing to equation (5) being more correct than equation (2).

TABLE XIV.

C.	I.	D.	P Pounds per Square Inch. Observed on Gauges.	Abbot's Formula.	Formula now Proposed.	REMARKS.
lb.		ft.				
500	...	30	18,368	$\alpha=130$ deg. 8,821	$\beta=+40$ deg. 16,791	Gun-cotton. English experiments, 1873. Crushers fixed in shells, buoyed from the surface.
500	...	38	14,784	$\alpha=143$ deg. 6,509	$\beta=+53$ deg. 13,494	
500	...	45	12,096	$\alpha=150$ deg. 5,210	$\beta=+60$ deg. 11,460	I = 100 considered a truer value than 87 for gun-cotton.
500	...	60	8,512	$\alpha=130$ deg. 3,344	$\beta=+40$ deg. 8,224	
500	...	23	15,792	$\alpha=90$ deg. 11,691	$\beta=0$ deg. 20,543	The gauges in these four were fixed to sinkers on the ground. Blow apparently deflected by the bottom.
500	...	30	*	$\alpha=90$ deg. 8,060	$\beta=0$ deg. 15,420	
500	...	37	9,184	$\alpha=90$ deg. 6,010	$\beta=0$ deg. 12,405	
500	...	44	7,840	$\alpha=90$ deg. 4,716	$\beta=0$ deg. 10,330	

* Rejected as a bad result.

An inspection of the analysis of the Oberon experiments and of the pressures now calculated by the above formula shows that a modern ironclad will receive a fatal injury if she be situated so that her outer skin occupies the position where crusher gauges should record a pressure P of about 12,000 lb. ($5\frac{1}{3}$ tons) on the square inch.

It will consequently be instructive to calculate the charges of different explosives which are required to produce this result at different distances.

Now,
$$P=\frac{9\,C\,I}{D}\left(1+\frac{25}{D^2}\right)$$
in horizontal plane. Consequently
$$C=\frac{P\,D}{9\,I}\left(\frac{D^2}{D^2+25}\right)$$
from which the following values for the charge in pounds are found for the different explosives at the distances given in the Table :

TABLE XV.

Distances in feet ...	2.5	5	10	20	30	40	50	
	Charges in pounds when P = 12,000.							I =
Description of explosive.	lb.	lb.	lb.	lb.	lb.	lb.	lb.	
Blasting gelatine ...	4.7	23.5	75	177	274	369	465	142
Forcite ,, ...	5.1	25	80	188	293	395	496	133
Gelatine dynamite ...	5.4	27	87	196	316	427	537	123
Dynamite and gun-cotton	6.6	33	107	251	389	525	660	100
Gunpowder	26.4	132	428	1004	1556	2100	2640	25

These results are plotted on the first diagram, which is useful for quickly finding the charge of either of the named explosives which is required to give a fatal blow to a modern ironclad at any intermediate distance.

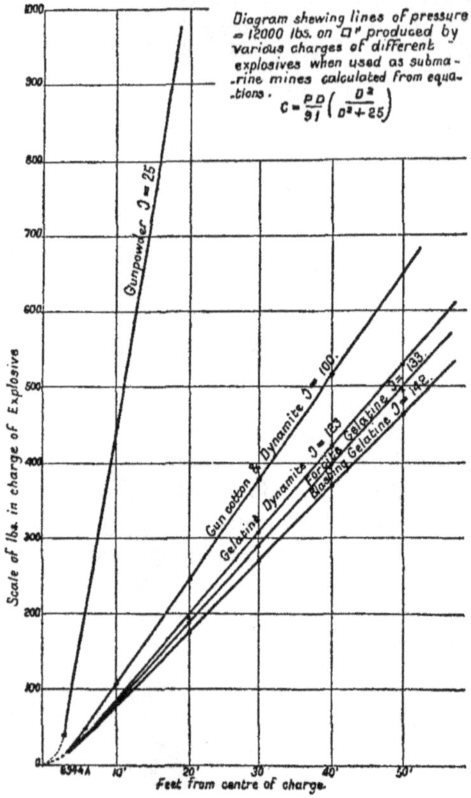

A large number of experiments were made with dynamite at Willet's Point, and it will be convenient to compare graphically the dynamite curves for different pressures as calculated from General Abbot's formula for P, which was based upon these experiments. The curves are shown on the second diagram, and they can be adapted to other explosives whose relative intensity of action is known, by using the simple proportion

$$P : P' :: I : I',$$

I for dynamite being 100 and I^1 the relative intensity of another explosive.

In a similar manner each curve can be adapted to another explosive.

Thus to adapt the curve marked 12,000 to blasting gelatine we have 12,000 : x :: 100 : 142. Hence $x = 17{,}040$, or the 12,000 lb. curve for dynamite becomes the 17,000 lb. curve for blasting gelatine, and similarly for the other curves on the diagram.

An examination of these curves will show that the charges of dynamite and of blasting gelatine required to give a pressure P = 6000 at various distances, are as follows, when P = 6000 as calculated by Abbot's formula, such pressure being considered by him as sufficient to fatally injure a man-of-war as strong as the Hercules :

TABLE XVI.

Distance D ... ... ft.	5	10	20	30	40
Blasting gelatine ... ... lb.	4	17	67	160	311
Dynamite ... ... ,,	7½	32	127	321	587

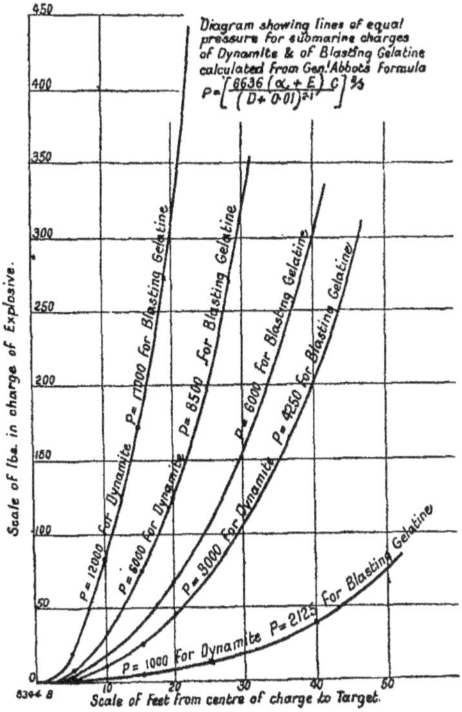

Taking into consideration the nature of an ironclad's double bottom, the charges for small distances derived from General Abbot's formula appear to be incapable of producing the damage required,

When $P = 12{,}000$ as calculated by the author's formula, such pressure being considered by him as necessary to insure a fatal injury to a modern man-of-war, we have :

TABLE XVII.

Distance D ... ... ft.	5	10	20	30	40
Blasting gelatine ... ... lb.	23½	75	177	274	369
Dynamite ... ... ,,	33	107	251	389	525

These charges are certainly more in accordance with practice, and with the results arrived at by numerous experiments.

CHAPTER IV.

EXAMINATION OF DIFFERENT EXPLOSIVES.

THE more important of the high explosives will now be briefly alluded to, and those of them which appear to be best suited for submarine work will be discussed minutely afterwards.

Dynamite.—I = 100. I being the relative intensity of action as compared with equal weights of other explosives. Composition = 75 per cent. of nitro-glycerine, 25 per cent. of keiselguhr. This explosive possesses so many advantages that it has been adopted by the American Government for their submarine mines, but it is now contemplated to employ one of the newer explosives, probably blasting gelatine. Dynamite will be examined again.

Gun-Cotton.—I = 100.* Abel's compressed gun-cotton or nitro-cellulose possesses so many advantages that it has been adopted by our Government and by several of the European Governments for submarine mining. It will be examined again.

Dualin.—I = 111.† Composition = 80 per cent. nitro-glycerine and 20 per cent. nitro-cellulose. Best form believed to consist of nitro-glycerine absorbed by Schultze's powder. In spite of its great power it is not well suited for submarine work, because it is "dangerous when frozen, and when saturated" (with water) "loses half its normal strength" (Abbot). It will not therefore be examined further.

Lithofracteur (or Rendrock).—I = 94. Composition = 40 per cent. nitro-glycerine and 40 per cent. sodium or potassium nitrate, 13 per cent. cellulose, and 7 per cent. paraffin. The principal absorbent being soluble renders this explosive unsuitable for submarine work. Other forms of rendrock, containing some more and some less nitro-glycerine, are equally open to this objection; and those with a larger percentage of nitro-glycerine are too moist to be safe for general use. Moreover, the salts used as absorbents being deliquescent, cause exudation of nitro-glycerine when the explosive is exposed to damp during storage.

* English experiments make I = 100, or slightly > 100.

† Somewhat similar to Abel's glyoxilin invented 1867.

Giant Powder.—I = 83. Composition = 36 per cent. nitro-glycerine and 48 per cent. sodium or potassium nitrate, 8 per cent. sulphur, 8 per cent. resin, coal, or charcoal. It is weaker than rendrock, and possesses similar disadvantages.

Vulcan Powder.—I = 82. Composition = 35 per cent. nitro-glycerine and 48 per cent. sodium nitrate, 10 per cent. charcoal, 7 per cent. sulphur. It is weaker than rendrock or than giant powder, and is open to the same objections.

Mica Powder.—I = 83. Composition = 52 per cent. nitro-glycerine and 48 per cent. powdered mica. Its strength as compared with other explosives of a similar character, is low.

Nitro-Glycerine.—I = 81. This value for intensity of action under water, which was verified by repeated experiments in America, was most unexpected. After laborious and careful investigation, General Abbot considers that nitro-glycerine is too quick in its action for submarine mining. He fully acknowledges that pure nitro-glycerine is more powerful than dynamite for rock blasting. But water is slightly compressible; and in order to obtain the best results, he thinks that a certain minute fraction of time is required in the development of the full force of the explosion. If this explanation be correct it may account for the superior power of wet over dry gun-cotton when used under water. It also explains the high coefficient given to blasting gelatine, which "is less quick and violent in its action than dynamite, although stronger." (*Vide* circular of manufacturers.)

Hercules Powder. I = 106. Composition = 77 per cent. nitro-glycerine and 20 per cent. magnesium carbonate, 2 per cent. cellulose, 1 per cent. sodium nitrate. Unsuitable for submarine mining, because the absorbents are soluble.

Electric Powder.—I = 69. Composition = 33 per cent. nitro-glycerine, and rest unknown. Weak, as compared with other nitro-glycerine explosives.

Designolle Powder.—I = 68.—Composition = 50 per cent. potassium picrate, 50 per cent. potassium nitrate. Dangerous, sensitive to friction, weak.

Brugere or Picric Powder.—I = 80. Composition = 50 per cent. ammonium picrate, 50 per cent. potassium nitrate. Safe, but weak.

Tonite.—I = 85. Composition = 52.5 per cent. of gun-cotton, 47.5 per cent. of nitrate of baryta. There are two varieties of this explosive, one dry, in compacted cartridges, the other damp, in bulk. It will be examined again.

Explosive Gelatine, 1881.—I = 117. Composition = 89 per cent. nitro-glycerine and 7 per cent. nitro-cotton, 4 per cent. camphor.

Blasting Gelatine, 1884.—I = 142. Composition = 92 per cent. nitro-glycerine and 8 per cent. nitro-cotton. This explosive is probably the best for submarine mining, and will be examined minutely presently.

Atlas Powder (A).—I = 100. Composition = 75 per cent. nitro-glycerine and 21 per cent. wood fibre, 2 per cent. magnesium carbonate, 2 per cent. sodium nitrate.

Atlas Powder (B).—I = 99. Composition = 50 per cent. nitro-glycerine and 34 per cent. sodium nitrate, 14 per cent. wood fibre, 2 per cent. magnesium carbonate.

As regards A, it would seem to possess no advantage over ordinary dynamite, and the 2 per cent. of sodium nitrate is objectionable, as it is deliquescent. As regards B, it would be a bad explosive for submarine mining, on account of the large amount of sodium nitrate. The power developed by this explosive is high, considering the percentage of nitro-glycerine.

Judson Powder (5).—1 = 78. Composition = 17.5 nitro-glycerine and rest unknown.

Judson Powder (3 F).—I = 62. Composition = 20 per cent. of nitro-glycerine and 53.9 per cent. sodium nitrate, 13.5 per cent. sulphur, 12.6 per cent. powdered coal, cannel.

These are very powerful considering the small amount of nitro-glycerine. Still more wonderful are the results obtained from the following grade of Judson powder, in which only 5 per cent. of nitro-glycerine is employed, and which costs about the same as common blasting gunpowder.

Judson Powder (C M)—I = 44. Composition = 5 per cent. nitro-glycerine and 64 per cent. sodium nitrate, 16 per cent. sulphur, 15 per cent. powdered cannel coal. This powder does not explode when struck by a bullet, nor when fired by a match, and although weak for submarine mining it would appear to be eminently suited for military land mining and work in the field. This grade of Judson's powder would perhaps be more properly classified under the heading of explosive mixtures, both on account of its composition and its strength, which is not sufficient to enable us to adapt it to Abbot's equations in which E appears.

Rackarock.—I = 88. Composition = 77.7 per cent. potassium chlorate, 22.2 per cent. nitro-benzole (insoluble liquid); also made in other proportions and with various modifications in the ingredients.

This explosive is one of the Sprengel class "which are non-explosive during their manufacture, storage, and transport." Their peculiarity is that the ingredients are kept separated until just before use. Until mixed, they cannot be exploded. When mixed a very powerful explosive is produced. A 3 oz. primer of tonite or of gun-cotton is required to detonate rackarock. The mixture is said to be stable, but the experiments made with it at Willet's Point indicate that its explosive action is not constant, probably due to the rough method in which the two ingredients are mixed. It appears to be inferior for submarine mining work to gelatine dynamite, to blasting gelatine, to forcite gelatine, gun-cotton, and dynamite.

Forcite Gelatine.—I = 133. Composition = 95 per cent. nitro-glycerine and 5 per cent. cellulose (un-nitrated). This explosive is claimed to possess certain advantages over blasting gelatine, which will be examined presently. As regards power for subaqueous work it is nearly as high as "blasting gelatine."

Gelatine Dynamite (No. 1).—I = 123. Composition = 65 per cent. of A and 35 per cent. of B. A = 97.5 per cent. nitro-glycerine and 2.5 per cent. soluble gun-cotton; B = 75 per cent. potassium nitrate, 24 per cent. cellulose, 1 per cent. soda.

This explosive does not appear to have been examined by General Abbot. The value for I, as given to the author by the manufacturers = 123, blasting gelatine at the same time being given at 153, but these values were not based on subaqueous crusher gauge experiments. General Abbot's value for I for blasting gelatine = 142, and the above value for I is found from the proportion 153 : 142 : : 132 : 123 nearly. Gelatine dynamite (No. 1) appears to be a suitable explosive for submarine mining, and will be examined again.

Gelignite.—I = 102, specific gravity = 1.5. Composition = 56.5 per cent. nitro-glycerine and 3.5 per cent. nitro-cotton, 8 per cent. wood meal, 32 per cent. potassium nitrate. The value for I given by the manufacturers (Nobel's Explosive Company, Limited) is 110, and is not based on crusher gauge experiments under water. The value for I given above is found in the same way as that for gelatine dynamite. Gelignite is but slightly more powerful than No. 1 dynamite, and as it possesses 32 per cent. of a soluble salt its employment in submarine mining cannot be recommended.

Melenite.—Composition (?). Very little as yet is known of this explosive, but it is believed to consist of fused picric acid in granules agglutinated with tri-nitro-cellulose dissolved in ether. Picric acid naturally takes the form of pure white crystals; but when these are fused and cast (a rather dangerous operation), it resembles beeswax. The intensity of action of this explosive is probably much exaggerated. The French Government is manufacturing large quantities. Until more is known about it the actual value cannot be compared with that of other explosives. Reports are conflicting and contradictory.

Roburite.—The intensity of action of this explosive under water is not known, but it has been adopted for use by Mr. Lay in his locomotive torpedo, and it is therefore in all probability a powerful submarine explosive.

The product of its gas volume and units of heat = 1,150,000, dynamite No. 1 being 950,000 (*vide* ENGINEERING, page 532, 8th November, 1887). From this it would appear that its relative power is 121 to 100 for dynamite and to 142 for blasting gelatine.

Experiments at Chatham have been carried out by the Royal Engineers, which show that its intensity of action when used for land purposes is distinctly inferior to gun-cotton. Carl Roth's specification, No. 9166, July 14, 1886, should be studied in connection with this new German explosive. It is one of the Sprengel class, like rackarock, but the ingredients are each solid. The patentee claims the process of producing explosives by the mixture of substances rich in oxygen, such as potassium nitrate, with a compound obtained from coal tar or from fractional products of the same, by incorporating therewith both chlorine and nitrogen. Six examples are given for producing such a compound, one of the most successful being thus made: 12 lb. of nitric acid (1.45 specific gravity) are heated with 4 lb. sodium chloride for one hour to 60 deg. Cent., and then cooled; 2 lb. naphthaline are then added in small portions to the mixture, and towards the conclusion of the reaction the whole is gently heated. A reddish mass separates out, and is freed from salt by washing. It is then digested for several hours with a mixture of three parts nitric acid (specific gravity 1.52) to six parts of sulphuric acid (concentrated). A brownish yellow crystalline substance is produced, which, after washing, &c., has a specific gravity of 1.4. By mixing one part of same with two parts of potassium nitrate a very powerful explosive is obtained. It is claimed "that the chlorine exerts a loosening effect on the atomic groups containing the nitrogen, and accordingly enables the said groups to react more readily on the oxygen yielding substances."

Roburite cannot be exploded by a blow or by friction. A red-hot poker can be put into a mass of it with impunity, and if placed in a smith's fire it burns slowly. Its explosion when effected by a strong detonating fuze proves it to be a powerful detonating mixture. A new departure in the manufacture of detonating explosives has been arrived at, but roburite does not appear to be well adapted for submarine mining, as its power is affected by damp. Its power can, however, be restored by drying.

The following Table gives the values of E for Abbot's formula for each explosive mentioned:

TABLE XVIII.

Explosives.	Intensity of Action under Water.	Value for E.
Dynamite	100	186*
Gun-cotton	87	135*A
"	100	E
Dualin	111	232
Lithofracteur or rendrock	94	160
Giant powder	83	120
Vulcan "	82	114
Mica "	83	119
Nitro-glycerine	81	111
Hercules powder	106	211
Electric "	69	67
Designolle "	68	65
Brugere or picric powder	80	110
Tonite	85	126
Explosive gelatine, 1881	117	259*
Blasting gelatine, 1884 ...	142	375*
Atlas powder (A)... ...	100	186
" " (B)... ...	99	183
Judson " (5) ...	78	100
" " (3 F) ...	62	45
" " (C M) ...	44	
Rackarock	88	140
Forcite gelatine	133	333*
Gelatine dynamite (No. 1)	123	254
" " (No. 2)	?	?
Gelignite	102	192
Roburite	?	?

REMARKS.—(A) Abbot. (E) English experiments. Those marked with a * are specially applicable for submarine work.

BEST EXPLOSIVES FOR SUBMARINE MINING.

The foregoing descriptions of the best-known high explosives show that dynamite, gun-cotton, gelatine dynamite, blasting gelatine, and forcite gelatine are the most suitable for submarine work. These explosives will now be examined in detail.

Dynamite.—I = 100, specific gravity = 1.6; has been before the public so many years and is so well known that it is not necessary to describe it. When slowly heated to 420 deg. Fahr. it is liable to explode with great violence. If frozen it should be thawed in a vessel jacketted with water at a temperature not exceeding 130 deg. Fahr. It is manufactured both in Europe and America, and is sold at a reasonable price—about 1s. 5d. a pound. It is powerful, easily detonated (too easily), it only loses 6 per cent. of its power when the charge is drowned by water, but General Abbot declares that the water does not cause exudation of nitro-glycerine if the charge be in the granular form and not in the form of compacted cartridges; it can be detonated when in the frozen state if granular and not in cartridges; it remains in good order after long storage, but it freezes at 40 deg. Fahr., and should be thawed before it is used; it is quite safe when handled with reasonable care; when used in the granulated state the cases can be loaded through a small hole; it does not vary with different samples, and is on the whole a trustworthy explosive. When using dynamite the printed instructions and cautions should be carefully followed. Its power is now, however, outmatched by blasting gelatine and forcite gelatine per unit of weight, and as these explosives are not equally open to the objection of producing the nitro-glycerine headaches to those who manipulate them, and are not so easily detonated, and therefore not so liable to sympathetic detonation, or to accidental explosion, especially when frozen, they are considered to be superior in many important particulars and inferior in none. Dynamite was chosen for the service explosive for submarine mining in the United States, but it is now understood that blasting gelatine may take its place for that service.

Gun-Cotton.—I = 100. This also has been a long time before the public, and its chief characteristics and manufacture are well known. It was discovered by the German Professor Schönbein, and the Austrians made many costly experiments with a view to introduce it as a war explosive. It was manufactured by them in a fibrous form and plaited into yarns, but the chemical and mechanical methods pursued did not free it from acid impurities. Such gun-cotton may become dangerous after storage. In the English method of manufacture the impurities are more thoroughly removed. Invented by a celebrated chemist who has made the study of explosives his speciality, and who has been the Government adviser for a great number of years, it has been developed under peculiar advantages and has been employed in the numerous experiments due to the evolution of torpedo warfare in this country.

The great safety with which in the wet state it can be stored and manipulated, and the important fact that it can be and is employed in the wet state as an explosive, constitute its chief merits. The fact that wet gun-cotton can be detonated was one of several important discoveries made by one of Sir Frederick Abel's assistant, the late Mr. E. O. Brown. Gun-cotton, when wet, is peculiarly insensitive to detonation, and consequently to sympathetic explosion when neighbouring charges are fired. This insensitivity is a great safeguard against acccidents of all kinds. When wet it can, like wood, be sawn or cut into any desired shape.

Its chief defect at the present date is want of strength (when used under water) per unit of weight or of cost as compared with other high explosives invented more recently. Also it is somewhat difficult and costly to manufacture in a pure and perfect condition, and the principal output is consequently confined to the Government establishments in those countries which have adopted it as a war store both for the land and sea services. Even when so made with the utmost care, its continued maximum efficiency after lengthened storage has never been attained except when stored dry. For this reason, and because it is generally stored wet, large quantities are not kept in reserve, and during a time of war it is therefore improbable that a sufficient amount could be manufactured to meet all requirements. When stored wet, it gradually becomes spongy from frequent wetting. This can be obviated by storing it between end plates firmly braced together by screw bolts as suggested by the author in January, 1875, but which has only been partially and imperfectly done in the mines of the Royal Navy. Sir Frederick Abel has always laid great stress on the necessity for retaining the density of the gun-cotton as issued from the manufactory, but General Abbot's experiments with gun-cotton appear to indicate that equal effects are produced under water from equal weights, whether the gun-cotton be in slabs or in the more bulky granulated form.

In buoyant mines the present practice is to employ an air space round the charge, and it would therefore appear to be immaterial whether the gun-cotton take the form of a solid mass or of a larger charge of the granulated material, so long as equal weights are inserted.

The amount of water usually added to dry gun-cotton, to wet it, is 25 per cent. Thus a charge of 125 lb. of wet gun-cotton contains 100 lb. of actual gun-cotton.

The size of the slabs used for the submarine mining service in England is 6⅛ in. by 6⅛ in. by 1¾ in., and each weighs about 2½ lb. when dry, the compression in manufacture being perpendicular to the larger surfaces, and the cleavage afterwards being parallel thereto. The above figures show that a cubic foot of English gun-cotton weighs about 66 lb. dry and 82½ lb. wet.

Gun-cotton is also made in Russia in cylinders 15 in. in diameter and 4½ in. high, weighing, when dry, 25 lb., which is at the rate of 54¼ lb. per cubic foot. The French gun-cotton is formed into slabs 4¾ in. × 4¾ in. × 1.6 in., weighing 22 oz., which gives the same specific gravity as the English gun-cotton.

As regards the relative intensity of action of gun-cotton when detonated under water, it appears that quick detonation in the open air affords no reliable measure of the force obtained for damaging a ship's bottom.

Sir Frederick Abel's experiments with small charges in air against iron plates a short distance off, although interesting, do not appear to lead to any useful data for submarine work. Also his experiments with small charges in bore-holes, although useful for rock blasting, are not for the most part applicable to submarine mining. For such work it is absolutely necessary to record, collate, and examine a great number of carefully conducted trials under water.

The low figure of efficiency for nitro-glycerine when used under water in its undiluted state forms a striking example, and shows how misleading may be the results of experiments in air when applied to subaqueous explosions.

As before stated, the American coefficients for relative intensity of action are principally based on a few experiments with small charges of each explosive, and the coefficient for gun-cotton, viz., 87 per cent. of that for dynamite, differs considerably from that which was obtained by experiments in England, where it was shown to be rather superior to No. 1 dynamite. Either the gun-cotton used in America was of inferior quality, or the dynamite used in England was inferior to that tried at Willet's Point.

Tonite.—I = 85. Composition = 52.5 per cent. gun-cotton, 47.5 per cent. of barium nitrate. Specific gravity = 1.28. This explosive is made at Faversham, in England, by the Cotton Powder Company, and also in California, U.S., under assigned patents. The English railroads carry tonite on the same conditions as gunpowder, but refuse to carry dynamite or compressed gun-cotton. Dry tonite is made up into candle-shaped cartridges covered with paper and waterproofed, usually with paraffin. Some of these cartridges are perforated to take a detonator, and are then called primers.

Wet tonite contains about 18 per cent. of added water, *i.e.*, 118 lb. of wet tonite contains 100 lb. of tonite. It is granular, uncompressed, and is taken from the incorporating mill before going to the press-room.

The experiments at Willet's Point indicate that equal amounts of tonite, whether dry in compacted cartridges, or wet in the uncompressed granular form, produce equal effects by submarine explosion.

Explosive Gelatine, 1881.—I = 117, specific gravity = 1.54. Composition = 89 per cent. nitro-glycerine; 7 per cent. nitro-cotton; 4 per cent. camphor. In 1867, Professor Abel combined nitro-glycerine and gun-cotton to form what he termed glyoxilin. He used tri-nitro cellulose. The explosive was practically a mechanical mixture. The percentage of nitro-glycerine was considerably less than in dynamite. Nobel afterwards found that when a lower product of the nitration of gun-cotton, viz., collodion or soluble gun-cotton (Abbot calls it nitro-cotton), be used in certain proportion in place of tri-nitro cellulose, there is a change, and the result has "almost the character of a compound" (Abel). By macerating from 10 to 7 per cent. of soluble gun-cotton with 90 to 93 per cent. of nitro-glycerine, a yellow, plastic, gummy jelly results, from which neither nitro-glycerine nor gun-cotton can be easily separated.

The addition of 4 per cent. of camphor makes explosive gelatine very insensitive to detonation from shock so long as it remains unfrozen. A rifle bullet at 100 yards' range striking a naked slab 3 in. thick, and flattening itself on an iron plate against which the slab rests, has failed to ignite the explosive.

Lengthened submersion in water causes little or no exudation of nitro-glycerine. It flames when ignited like dry gun-cotton or dynamite. It becomes soft and somewhat greasy at 140 deg. Fahr., and it freezes at about 40 deg. Fahr. It can be cut with a knife. Its specific gravity is 1.54. The presence of the camphor prevents it from detonating, as it otherwise does when heated slowly to 400 deg. Fahr. When camphorated it burns with sparks at about 570 deg. Fahr. To insure its detonation the Austrians employ a special primer made of 60 per cent. nitro-glycerine and 40 per cent. nitro-hydro-cellulose (Jekyll. Royal Engineer Corps papers). The latter is formed by first treating cotton with sulphuric and afterwards with nitric acid. When mixed with the nitro-glycerine, a white soapy substance is formed, 20 grammes of which suffice to detonate a charge of explosive gelatine with certainty. The intensity of action $I = 117$ was obtained from the American experiments in 1881, in which the explosive was of inferior quality, subsequent long storage of a portion of it proving that the nitro-cotton was impure. Moreover, the gun-cotton or dynamite priming charges employed sometimes failed altogether to detonate the charges of gelatine, which must have been bad.

Blasting Gelatine, 1884.—$I = 142$. Composition = 92 per cent. nitro-glycerine and 8 per cent. collodion gun-cotton, specific gravity = 1.53 to 1.55. The sample (2000 lb.) for Abbot's experiments was supplied, as to the trade, from Scotland without any added camphor. The makers, Nobel's Explosive Company, state that it can be added if desired, as follows: "Warm gently by means of water at 60 deg. Cent., and when the gelatine attains a soft plastic state, with a temperature of 40 deg. Cent. in the mass, 5 per cent. of camphor dissolved in alcohol may be added and completely incorporated with the hand to form a homogeneous mass. The warming is best done in a copper basin surrounded with water at 60 deg. Cent.; 40 lb. or 50 lb. may be camphorated at a time in the above manner."

The experiments made by General Abbot, from whose records the above value of I is taken, were carried out with the uncamphorated gelatine, and most of the charges were detonated with a service (American) fuze, viz., a copper capsule containing 24 grains of fulminating mercury. Several shots were also fired with a 3-oz. tonite primer, and the results proved that full effects were produced with the fuze alone.

Experiments were made to test it for sympathetic explosion, the gelatine being placed in thin rubber bags at various distances from a primary charge of 1 lb. of dynamite. At 5 ft. explosion occurred, but at 5 ft. 9 in. and over no explosion occurred. It would have been more satisfactory if the primary charge had been 1 lb. of the gelatine. Naked charges hung against wooden boards were fired at by a rifle at a range of twenty paces. The gelatine blazed when struck, and on one occasion a small explosion occurred, throwing unignited fragments of the cartridges a few feet from the target. When lighted with a match it burns with an intense white flame.

"Blasting gelatine without camphor is most admirably suited to submarine mining, in so far as strength and ordinary physical properties are concerned. For military purposes on land a small percentage of camphor should not be omitted." (Abbot.)

This explosive, whether camphorated or not, may be kept immersed in water for a length of time without undergoing any important change. "It has consequently been proposed to render the storage of blasting gelatine and certain of its preparations comparatively safe by keeping them immersed in water till required for use." (Abel.) This is rather hard on blasting gelatine, for it implies that this precaution is necessary in order to make its storage only "comparatively safe." So far as present knowledge goes, this explosive can be stored dry as safely as gunpowder, "cool and dark storage" being secured whenever possible. (Abbot.) The deterioration in store which occurred in the early samples sent to America, were apparently due to impurities in the nitro-cotton, but this is now provided against by great care in the manufacture. Nevertheless, the difficulty of producing pure nitro-cotton is well known, and this leads us to forcite, in which it is avoided.

Forcite (No. 1 extra).—I = 133. Composition = 95 per cent. nitro-glycerine, and 5 per cent. cellulose.

Forcite (No. 1).—I = 124. Composition = 75 per cent. nitro-glycerine, 7 per cent. cellulose, and 18 per cent. nitre.

Forcite (No. 2).—I = 95. Composition = 40 per cent. nitro-glycerine and 60 per cent. explosive base (nature unknown).

Forcite (No. 3 C).—I = 88. Composition = 30 per cent. nitro-glycerine and 70 per cent. explosive base (nature unknown).

Specific gravity, No. 1 extra = 1.51, No. 1 = 1.6, No. 3 = 1.66, No. 3 C = 1.69.

They can all be detonated with a fuze containing 24 grains of fulminating mercury. No increased power was obtained when a larger priming charge = 3 oz. of tonite was employed. The explosive base employed in the lower grades is probably some combination of sodium nitrate with resin, coal dust, &c., mixed with cellulose and sometimes with dextrine. As regards sympathetic explosion, the highest grade acts similarly to explosive gelatine, exploding at 5 ft. from a primary charge of 1 lb. of dynamite under water. It will be remembered that dynamite itself explodes at 20 ft. under like conditions. "No. 1 extra" forcite—or forcite gelatine—contains no *gun-cotton* or *nitro-cotton*, but simply unnitrated cellulose, combined with nitro-glycerine. Cotton is treated alternately with acids and alkalies, as for paper stock, leaving pure cellulose which is reduced to a powder and then exposed to high-pressure steam in a closed vessel until it becomes a gelatinous mass. This can be stored for any length of time in water; 95 per cent. of nitro-glycerine can be incorporated with it to form a highly explosive jelly, very similar to explosive gelatine, both in its appearance and properties. It is claimed, and apparently with justice, that its manufacture is less costly than similar compositions of nitro-glycerine. Also that the nitro-glycerine is so completely incorporated that it cannot be separated even by the application of alcohol or sulphuric ether. Also that water has no action upon it, that it detonates with the greatest violence, that it burns away harmlessly in the open air. General Abbot concludes the report on a careful series of small charge experiments with forcite as follows: "These investigations indicate that forcite must be classed as one of the explosives worthy of serious consideration when it becomes necessary to defend our coasts with submarine mines. Its great strength is fully established; its permanency for long periods of time remains to be studied."

Inasmuch as pure cellulose is easier to manufacture than pure nitro-cellulose, it would appear that its permanency can be more easily assured than that of explosive or blasting gelatine. As regards their comparative intensities of action, the difference is in favour of Nobel's explosive. Forcite is the invention of a French chemist, M. John M. Lewin, who patented it in Belgium in November, 1880, but it is evidently a very close copy of blasting gelatine.

Gelatine Dynamite (No. 1).—I = 123, specific gravity = 1.55. (No. 2) I = not known. Composition (No. 1) = 65 per cent. A and 35 per cent.

B. Composition (No. 2) = 45 per cent. A and 55 per cent. B. A = 97.5 per cent. nitro-glycerine and 2.5 per cent. soluble gun-cotton, B = 75 per cent. potassium nitrate, 24 per cent. cellulose, 1 per cent. soda. These lower grades of explosive gelatine are manufactured and sold to compete with the various explosives in the market for blasting purposes. They have not been tried by General Abbot, and the value of I given was found as follows: The makers (Nobel's Explosive Company) state that its relative intensity of action is 132, but they also give blasting gelatine at 153. For submarine work blasting gelatine should be 142 according to Abbot, and reducing 132 in the same proportion, 123 is obtained.

The makers of the gelatine dynamites state that they are much more powerful than dynamite, more convenient to handle, and more economical, *i.e.*, a greater effect per unit of cost. The latter is open to doubt (see Table, page 54). Moreover, they are unaffected by water, and are less sensitive to detonation, and therefore to accidental explosion by a blow. They require to be detonated by special gelatine detonators supplied by the manufacturers. They freeze at 40 deg. Fahr. When frozen they should be carefully thawed by means of a water bath (water not over 130 deg. Fahr.) in accordance with the printed directions issued with them. They should never be exposed to a tropical sun.

Gelatine dynamite (No. 1) appears to be well adapted for submarine work.

Generally, all the nitro-glycerine compounds should not be kept for long periods at a higher temperature than 130 deg. Fahr. They can be tested in small quantities to 160 deg. Fahr., but any temperature over 140 deg. Fahr. is dangerous.

In conclusion, the efficiency of an explosive for submarine mining depends not only upon the intensity of action per unit of weight, but upon the intensity of action per unit of cost, and also per unit of space occupied.

The approximate cost and weight of each of the best explosives for submarine mining, as well as their relative intensities of action per unit of weight, cost, and space, are given in Table on next page. It will be seen that blasting gelatine heads the list in every case, although lower values are given to it than the manufacturers claim as its due. For ground mines gelatine dynamite and dynamite are nearly as economical as blasting gelatine, but they require larger and therefore more expensive mine cases.

For buoyant mines blasting gelatine is the best explosive.

When the high explosives cannot be obtained in sufficient quantities at a time of emergency, gunpowder can be used effectively, especially in the ground mines; a line is therefore added to the Table for gunpowder. When used every care must be taken to place it in strong cases, and to ignite it by means of lightning fuze coiled near the outside of the charge This causes the outside of the charge to be ignited first, forming an outer surface of gas at a high temperature, and protecting that portion of the charge which is ignited last from being drowned when the case is ruptured. In this manner it is probable that the whole of the gunpowder charge would be ignited and burnt. Such an arrangement is especially necessary when gunpowder mines are improvised with weak cases like barrels. It should be noted that owing to its low price, the efficiency of gunpowder per unit of cost is more than three times its efficiency per unit of weight, but the figure 85 somewhat overstates the matter, because a larger and therefore a more expensive mine case is required when gunpowder is used for submarine work.

Experiments of late years have been conducted almost entirely with gun-cotton in England and dynamite abroad. Gunpowder has been neglected. It is very desirable that some large ground charge experiments should be conducted with this useful and well-known explosive, and perhaps with some of the other forms of cheap explosive mixtures, especially Judson's powder.

TABLE XIX.—RELATIVE VALUES OF EXPLOSIVES FOR SUBMARINE MINING.

Order of Merit.	Description of Explosive.	Specific Gravity.	Weight of Cubic Foot in Pounds.	Cost per Pound in Pence.	Efficiency under Water.			REMARKS.
					Per Pound.	Per Cubic Foot.	Per £.	
1	Blasting gelatine	1.54	96.3	24	142	138	101	
2	Forcite gelatine ...	1.51	95.4	?	133	127	?	The manufacturing cost cannot be greater than that of blasting gelatine, and is probably less.
3	Gelatine dynamite	1.55	96.9	21	123	119	99	
4	Dynamite, No. 1.	1.6	100	17	100	100	100	
5	Gun-cotton, dry ...	1.06	66	27	100	66	63	
5	Gun-cotton, wet...	1.32	82.5	...	80	66	63	25 per cent. of added water.
6	Tonite	1.28	80	...	85	68		
7	Gunpowder ...	0.9	56	5	25	14	85	

CHAPTER V.

Considerations Guiding the Size and Nature of Mine Cases, &c.

All submarine mines can be classified as follows. Class A: Mines that are caused to explode when in contact with, or very close to a vessel's side or bottom. Class B: Mines that act at a greater distance.

All submarine mines can also be placed in two divisions. Division 1: Mines under control; these are electrical. Division 2: Mines not under control, whether they be electrical, mechanical, chemical, or some combination of these three.

Concerning Class A, it has already been calculated that 33 lb. of gun-cotton or dynamite properly placed form an ample charge to fatally injure a modern man-of-war, and this agrees with the experiments against the Oberon, which show that 33 lb. of gun-cotton contained in a case without any air space, and not enveloped with a wooden jacket, will break through the double skin of H.M.S. Hercules when exploded 4 ft. off it.*

The Employment of Wooden Jackets not Desirable.—The great loss of power produced by surrounding the mine case with a sheathing of wood, or other similar material such as cork, has already been noted. The employment of wooden or cork jackets to give buoyancy to mines cannot therefore be entertained for a moment.

Loss of Effect due to an Air Space in a Mine.—An equal loss of power is not, however, produced by employing an air space in a mine to give the required buoyancy. Experiments made in America to test the effect of such air space show that a certain loss of power is occasioned. Unfortunately the charges employed did not exceed 8 lb. of dynamite, and the cases were simply tin cylinders. These experiments demonstrated that with an air space not exceeding three times the volume of the charge no sensible effect on the intensity of action was observed; but when the air space was increased to five times the volume of the charge of dynamite, the mean pressure recorded fell from 8554 to 6038, a loss of nearly 30 per cent. in the intensity of action as recorded in crusher gauges placed 8 ft. from the centre of each charge.

* See foot-note page 59.

General Abbot remarks on the results: "The safe limit of void space for small charges fired under water lies, therefore, between three and five times that of the charge."

"It would appear altogether probable that as the size of the charge is increased, this limit increases also, and hence, that the void space necessary to give the requisite flotation to buoyant torpedoes does not lessen their destructive power. For ordinary mines of this character the void space is usually between two and three times that occupied by the charge."

Whatever may be the practice in America, such a small air space is inadequate when the cases are made of sufficient thickness to resist countermines, and of sufficient size to counteract by their buoyancy the sinking action produced by the side pressure of a strong tidal current acting on a mine secured to the bottom by a single mooring, as is usual in England. It is therefore to be regretted that no experiments have been made in England or elsewhere to test the effect of employing an air space, such as that employed in our service, and which is nearly twelve times the volume of the charge. It is possible that so large an air space may act very prejudicially, and warrant a different arrangement of charge, mine case, &c.

The Evils of Using Larger Charges than are Absolutely Necessary.—In all demolitions an engineer should endeavour to use charges, each of which is a thoroughly effective minimum, and this is especially necessary in buoyant submarine mines. Unnecessarily large charges in contact mines produce many evils, which act and react in a somewhat peculiar manner.

1. A large charge requires a large case to buoy it up even in dead water.

2. The large case opposes a greater resistance to tidal or other currents, and has to be still further increased when moored in such currents.

3. The large case must be made of thicker metal if intended to be of equal efficiency with the small case to resist the effects of neighbouring explosions.

4. These neighbouring explosions being produced by mines, like itself, which are unduly powerful, it must be made of still thicker metal in order to be safe from them.

5. This again requires the case to be still larger in order to buoy up the thicker metal.

6. If the mines be spaced further apart to obviate this, it does not provide a strong case to resist countermining by a foe.

7. There is a waste of money in explosives.

8. The mooring gear must be heavier than necessary.

9. The periods of time required to load and connect up the mines, and lay them and pick them up for repair, must all be longer than necessary, and this alone is a most important consideration even to a country that can afford to use the larger and more costly mines on the score (whether true or not) of increased efficiency.

Other considerations, all pointing in the same direction, could be mentioned, but enough has perhaps been said to show that the best contact mine is one that contains a thoroughly effective, but a minimum charge.

Existing Custom.—It is and always has been usual to place the apparatus for firing and testing an electro-contact mine, and the air space for buoying it, and the charge, in one case. It is perhaps better not to do so. By arranging the charge in a small case that will just contain it, the full force of the blow, uncushioned by any air space or wooden jacket, is transmitted direct to the vessel's side. The buoy can be placed over the mine and separated from it by a distance apportioned according to the charge employed. When a charge of 100 lb. of gun-cotton or dynamite, or anything approaching it, is used, the mine should be 8 ft. under the top of the buoy, and the electrical apparatus be placed in the latter. The actual distance of the exploding charge from the vessel will then be either something less than 8 ft. under her bottom, or the charge will be close to the ship's side, if not in actual contact with it. The term "contact mine" for such an arrangement may appear to be a misnomer, but it is a convenient one.

A comparison of the results of the Oberon experiments with those recently carried out at Portsmouth against H.M.S. Resistance, as recorded in the *Times*, appear to indicate that small charges act with greater effect when not in actual contact with a vessel's side, but slightly removed from it. There is then also a greater chance of making large holes through both skins of the double bottom, and a greater probability of drowning more than one compartment of the vessel, and of placing her out of action. If then a charge as large as 100 lb. be used, the above seems to be the most reasonable and effective arrangement.

For reasons already given, the employment of a smaller charge is advocated, and it then becomes necessary to bring it closer to the vessel. This can be done by suspending it about 4 ft. below the top of the buoy. Fig. 21 shows the usual arrangement now in vogue. Fig. 22 shows the modification now recommended if the large 100 lb. charge be retained. Fig. 23 shows the plan recommended when a smaller but sufficient charge is used. Fig. 24 shows the same arrangement in the

mines and rough handling. The required minima are: 1. Weight. 2. Resistance to moving water.

As regards shape the above maxima and minima are provided for best by the sphere. Buoyancy and weight being antagonistic, it is necessary to arrive at some decision concerning the strength which is required in practice. It may be accepted that a spherical case 3 ft. in diameter, made of $\frac{1}{4}$-in. steel, is strong enough, and as the strength to resist a collapsing pressure varies inversely with the square of the diameter, and directly as the square of the thickness, we can at once find the thickness of other spherical cases of the same material which shall possess equal strength. Thus, if

s = strength required,
t = thickness of steel in inches,
d = diameter of sphere in inches.

Then, as

$$S \propto \frac{t^2}{d^2} \text{ and is sufficient when}$$

$$t = \tfrac{1}{4} \text{ and } d = 36$$

$$\frac{t^2}{d^2} = \frac{(\frac{1}{4})^2}{(36)^2}$$

$$\therefore t = \frac{d}{144}.$$

The following Table shows the thickness of steel required for spherical cases or buoys of various dimensions, but of equal strength, to a 3-ft. case or buoy of $\frac{1}{4}$-in. steel:

TABLE XX.

Diameter of Sphere.	Area of Diametric Plane.	Thickness of Steel in 24ths in.	Weight of Shell.	Weight Salt Water Displaced	Buoyancy Empty.
ft.	sq. ft.	in.	lb.	lb.	lb.
1	0.78	2	11	32	21
1.5	1.76	3	34	115	81
2	3.14	4	84	269	185
2.5	4.9	5	160	525	365
3	7.06	6	283	903	620
3.5	9.6	7	440	1427	987
4	12.57	8	672	2144	1472
5	19.6	10	1352	4220	2870

The diagram of curves, giving similar information, is also of use. Actually, the weight of the case or buoy in air always exceeds that of the bare shell on account of the double thickness at the joints, and because the cases are generally strengthened by rings of ⊥ or L irons, and the mouth of the case or buoy by a ring of wrought iron, malleable cast iron, or steel. The weights of these additions must therefore be deducted from the buoyancy recorded in last column of Table XX.

As regards size, the available buoyancy, *i.e.*, that which remains after deducting all the weights which have to be supported, should be ample to prevent the system being drawn down by tidal currents below the limiting horizontal plane of action of the ship's bottom.

This can be insured with a small buoyancy if the system be moored on a span by two sinkers or anchors—one up, the other down stream. The system then remains in one position whether the tidal current run up or down stream, as well as during slack water.

Such an arrangement is well adapted for a mine fired by an observer or observers at a distance, but there is not the same necessity to keep a contact mine exactly in one place, and there are certain practical difficulties in laying mines on two moorings. The single mooring has

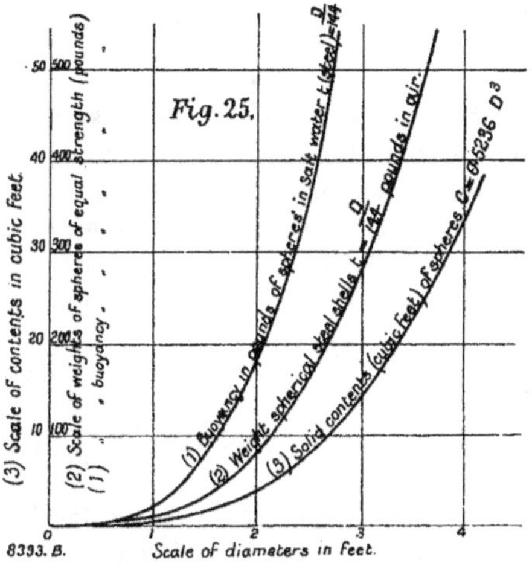

Fig. 25.

therefore been adopted for contact mines in Europe, but the following calculations will show how difficult it is to use such mines so moored in tidal currents. Assume that the charge of explosive is 100 lb., and that the weight of the circuit-closing apparatus with its metal envelope together with the weight in sea water of the mooring cable and mooring line, with shackles, attachment chains, &c., amount to 100 lb. more, total, 200 lb. deadweight.*

Let the depth be eleven fathoms at low water, draught of vessels to be four fathoms, rise and fall of tide two fathoms, and centre of mine always covered by one fathom, *i.e.*, at low-water slack. Evidently the mine must not be more than four fathoms below the surface just before and just after high water when the tidal current is running.

* If the service circuit-closer and mouthpiece were used, this deadweight would be considerably greater.

The conditions being plotted geometrically (see diagram, Fig. 26) it will be found that the angle a which the mooring line makes with the vertical is about 24 deg. If B denote the available buoyancy of the mine when loaded and moored, and if P denote the side pressure in pounds on the mine caused by the current, and P^1 the pressure on the mooring line and cable; then, taking moments round the sinker,

$$4.2\,B - 9\,P - 4.5\,P^1 = 0.$$

The resistance P in pounds offered by a sphere to salt water flowing past it with a velocity V in knots per hour is

$$P = 1.03\,V^2\,A,$$

where A is the area in square feet of the diametric plane (*vide* letter from the late Mr. W. Froude, extracts from which will be given later).

Also, the resistance offered by an upright cylinder under the same conditions is

$$P^1 = 2.85\,V^2\,A',$$

Assume that the mine is large enough to keep the mooring line nearly straight, and that the resistance offered by a long cylinder (length $= l$), when tilted, is approximately equal to that offered by a vertical cylinder whose length is $l \cos a$. The vertical component for the wire rope will be 60×0.9 ft., and as the electric cable has $\frac{1}{12}$ slack, its vertical component will be 65×0.9 ft. The diameter of the wire rope is $\frac{2}{3 \times 12}$ ft., and that of the electric cable $\frac{8}{10 \times 12}$ ft. Consequently

$$P^1 = 2.85\,V^2 \left(60 \times 0.9 \times \frac{2}{3 \times 12} + 65 \times 0.9 \times \frac{8}{10 \times 12} \right) = 19.66\,V^2.$$

[In place of taking moments round the sinker, the formula

$$B = V^2 \cot a\,(1.03\,A + 1.42\,A^1)$$

may be employed, and the same result obtained.]

Now $B = \frac{1}{4.2}(9\,P \times 4.5\,P^1)$

$= V^2\,(2.21\,A + 21.1).$

And when $V = 2$ knots per hour,

$B = 8.83\,A + 84.$

And when $V = 3$ knots per hour,

$B = 20\,A + 190.$

And when $V = 4$ knots per hour,

$B = 35.3\,A + 337.$

And when $V = 5$ knots per hour,

$B = 55.1\,A + 527.$

As before stated, however, the real buoyancy of the case must hold up 200 lb. in addition to its own weight. The actual buoyancy of the mine case, when empty, must therefore be:

For a 2-knot tide = 8.8 A + 284
,, 3 ,, = 20 A + 390
,, 4 ,, = 35.3 A + 537
,, 5 ,, = 55.1 A + 727

Referring to the table of buoyancy, &c., for steel spherical cases, whose thickness $t = \frac{D}{144}$, it will be found that the above equations are satisfied when D (the diameter of the case) is 2.4 ft. for a 2-knot tide, 2.8 ft. for 3 knots, 3.25 ft. for 4 knots, and 3.8 ft. for 5 knots.

These figures would have to be increased if the cases were manufactured in the manner practised in this country, viz., if several heavy iron strengthening rings are added to the interior of the cases. These rings weigh about 100 lb. in the 3-ft. cases, but their employment is entirely uncalled for in spherical cases, either as a manufacturing necessity or as an assistance to withstand the shock of countermines. The same amount of metal added to the thickness of the skin of the case is evidently a much better method of employing it. There is no difficulty in forming a lap joint at the junction of the two hemispheres.

But the question of available buoyancy requires further investigation before entering upon the methods of manufacture. The instance given was one in which no especial difficulties arose, yet the size of the case required for a 5-knot current, and even for a 4-knot current, was larger than those which are usually employed for contact mines, and every endeavour should be made to decrease their size for the reasons given already, on pages 56 and 57. But, in the first place, it will be well to note how utterly all contact mines must fail when the rise and fall of tide exceeds two fathoms or thereabouts, and when the method of single mooring is adhered to. In the last example, if the rise and fall of tide be three fathoms instead of two, and the other conditions remain as before, it is evident that when the tidal currents are flowing just before and just after high water (and they frequently run very hard at the first of the ebb out of a large land-locked harbour) the mines will be submerged five fathoms, and will consequently be out of the plane of action of a vessel's bottom drawing four fathoms or thereabouts. Assuming that the mines are only drawn down one fathom by the action of the current, double mooring would only have this advantage of one fathom; but it is very important in a large percentage of tidal harbours, where the rise and fall of tide exceeds two, but does not exceed three fathoms. When it exceeds three fathoms it is necessary to adopt some of various devices which will be explained hereafter. But the rise and fall may be moderate, and yet the tidal currents and the depth of water be considerably greater than in the instance examined. Such conditions are met with at the entrance of the Solent. Assume that the depth is twenty-one fathoms in place of eleven as before; then, other conditions remaining unaltered, it will be found by geometrical construction that when a single mooring is employed $a = 18$ deg., and that, taking moments round the sinker,

$$5.8\,B - 19\,P - 9.5\,P^1 = 0.$$

P^1 found as before $= 2.85\,V^2\,(6+7.8) = 49.33\,V^2$.

And P as before $= 1.03\,V^2\,A$.

Consequently, $B = \frac{1}{5.8}\,(19\,P + 9.5\,P^1)$

$= V^2\,(3.38 + 80.8)$.

And when $V = 2$ knots per hour,
$B = 13.5\,A + 323$.

And when $V = 3$ knots per hour,
$B = 30.2\,A + 727$.

And when $V = 4$ knots per hour,
$B = 54\,A + 1292$.

And when $V = 5$ knots per hour,
$B = 84.2\,A + 2020$.

But the actual buoyancy must hold up 50 lb. more cable and mooring line than in the last example, or a total of 250 lb. in addition to the weight of the case, and the above equations therefore become when so adjusted:

$B = 13.5\,A + 573$ for a 2-knot current.
$= 30.2\,A + 977$,, 3 ,,
$= 54\,A + 1542$,, 4 ,,
$= 84.2\,A + 2270$,, 5 ,,

Referring to the Table we find that these equations are satisfied when D (the diameter of the case) is 3.1 ft. for a 2 knot tide, 3.9 ft. for 3 knots, 4.9 ft. for 4 knots, and as this is too large a case for practical use, it is unnecessary to work out the figures for a 5-knot tide.

We will now apply a mine on double moorings to the last example, and note the results.

A glance at the diagram Fig. 26 shows that the mine can be submerged two fathoms at low water, if the rise and fall of tide be no greater than two fathoms. Such a mine is therefore less likely to be seen at low-water slack than a mine on single mooring which must, under the same conditions, rise one fathom higher at this time of tide, if moored to catch a four-fathom vessel at all times. Another and a very important practical advantage in favour of two moorings is also shown upon the same diagram. Before laying mines on single moorings, it is necessary to survey the waters to be mined with the greatest accuracy, and subsequently to lay the mines exactly in the positions where soundings have been taken. When the sea bottom is irregular it is most difficult to lay the mines at the correct level, and in spite of every precaution they often have to be raised and their positions shifted or their mooring lines altered in length.

When mines are moored each on a span between two sinkers, or between an anchor and a sinker, a very accurate survey of the waters to be mined is less essential, because the submersion of the mines can be rectified when laying the last sinker, and any irregularity of sea bottom is automatically allowed for in this process. Thus, the anchor A (Fig. 26) being laid first by means of its wire rope mooring line, the

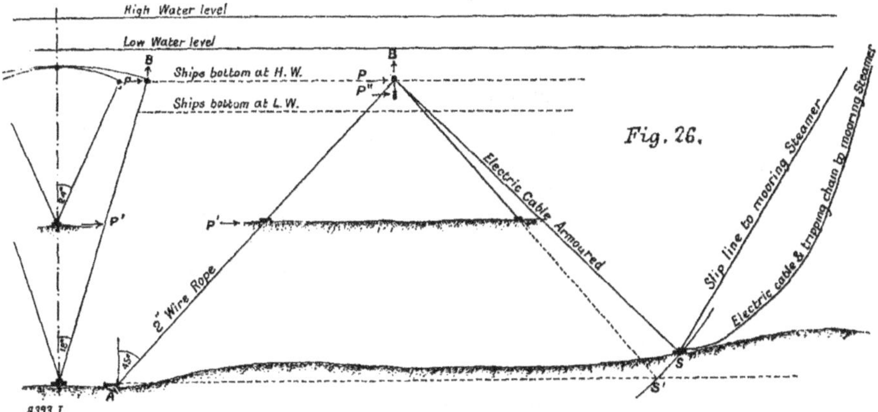

Diagram showing Mine on Single & Mine on Double mooring at depths of 21 & 11 fathoms.

mine is allowed to float on the surface. The mooring steamer then drops down with the current, keeping its head up stream, pointing on the position of the anchor, and paying out the electric cable. The sinker is now lowered by a line until it touches the bottom, and the electric cable is slacked off at the same time. A boat attends at the mine, and attaches a measuring line to it. The steamer then adjusts the position of the sinker until the mine is at the required submersion, which of course varies with the time of tide. The boat hails when the desired result is obtained, the end of the line is then stoppered to the cable, which is paid out and taken by the steamer to any desired place. If the bottom be level the sinker will be placed at S', the span being equally divided. But if the bottom be irregular, as shown in diagram, the sinker will be placed at S and the two mooring lines will take slightly different tilts. To facilitate this operation a heavy sinker should be used at S, and the angle between the lines at the mine should not be much less than a right angle. A good spread in the span also decreases the necessary buoyancy of the mine.

The following calculations give the size, &c., of a detached circuit-closer moored on a span as shown, and supporting 4 ft. below it a small mine containing 36 lb. of blasting gelatine in a case weighing 10 lb. in salt water, or 46 lb. when loaded.* The circuit-closing apparatus with envelope to weigh 20 lb. The buoy to be spherical, and of steel, with a thickness equal to $\frac{D}{144}$. The mine to be moored by the electric cable on one side and a steel wire rope on the other side to two anchors or sinkers up and down stream. The cable to be armoured, and to weigh 25.5 cwt. per knot in air, and 13 cwt. per knot, or 1.45 lb. per fathom, in salt water. Its diameter to be about 0.8 in., and its breaking strength 5 tons. The wire rope to be 0.6 in. in diameter, to weigh 3 lb. in air, and 2.5 lb. in salt water per fathom; also to possess a breaking strength of about 5 tons.

When moored in 21 fathoms, the buoy will therefore have to support $27\ (1.45+2.5)=108$ lb. of cable and rope, as well as 46 lb. for the small loaded mine, or 154 lb. in all, besides its own weight. Also, the area of the diametric plane opposing the current is for the cable $= 27 \times 6 \times \frac{8}{10 \times 12} \cos 45$, and for the rope $27 \times 6 \times \frac{6}{10 \times 12} \cos 45$. Consequently the total vertical area given by the cable and rope $=13.4$ square feet, and the small mine case offers half a square foot.

Proceeding as in previous examples:

$$19\ B - 19\ P - 18\ P^{11} - 9.5\ P^{1} = 0.$$

When B = available buoyancy of sphere,
P = side pressure on same,
P^{11} = same on small mine case,
P^{1} = same on cable and rope.

$$\therefore B = P + \frac{18}{19} P^{11} + \tfrac{1}{2} P^{1}.$$

But $P = 1.03\ V^2 A$.
When A = diametric area of the sphere,
and $P^{11} = 2.85\ V^2\ \tfrac{1}{2}$,
and $P^{1} = 2.85\ V^2\ 13.4$.
Hence $B = V^2\ (1.03\ A + 23.25)$.
When V = 2 knots per hour,
$B = 4.12\ A + 93$.
When V = 3 knots per hour,
$B = 2.27\ A + 209.25$.
When V = 4 knots per hour,
$B = 16.48\ A + 372$.
When V = 5 knots per hour,
$B = 25.75\ A + 581.25$.

But, as already stated, the actual buoyancy of the sphere must

* This calculation should be slightly modified (see foot-note p. 59), but the results are not materially affected.

be greater than the above by 154 lb. The buoyancy so altered then

$$= 4.12\,A + 247 \text{ for a 2-knot tide.}$$
$$= 9.27\,A + 368 \;\;,, \;\; 3 \;\;,,$$
$$= 16.48\,A + 526 \;\;,, \;\; 4 \;\;,,$$
$$= 25.75\,A + 735 \;\;,, \;\; 5 \;\;,,$$

Referring to the Table we find that these equations are satisfied when D (the diameter of the buoy or detached circuit-closer) is 2.3 ft. for a 2-knot tide, 2.6 ft. for 3 knots, 3.1 ft. for 4 knots, and 3.5 ft. for 5 knots.

Comparing these results with those calculated for the service arrangement under like conditions (depth of water, draught of vessel, &c.), we find as follows:

TABLE XXI.

Current in Knots per Hour.	Size of Sphere Required, Diameter, Feet.	
	Single Mooring.	Double Moorings.
2	3.1	2.3
3	3.9	2.6
4	4.9	3.1
5	Not calculated	3.5

From this we see that a service case, &c., cannot be used in 21 fathoms when current exceeds 2 knots. But with two sinkers and everything made as light as possible, yet possessing sufficient strength, a 3-ft. sphere will act effectively in the same depth in a current of nearly 4 knots. Also a 2½-ft. sphere in 21 fathoms and 3 knots. Also a similar calculation will prove that a 2½-ft. sphere will act efficiently in a 3½-knot current, and a depth of 11 fathoms. Evidently two moorings should be used with each mine exposed to a strong tidal current.

The Late Mr. W. Froude's Formulæ for Side Pressure.—It will now be convenient to quote from the valuable letter written by the late Mr. W. Froude, May 18, 1876, and already referred to: "Sir,—I have read with interest and attention your letter of the 16th and its inclosures relating to the forces to which a submarine buoyant torpedo is subject when moored in a tideway or current of known velocity, and I will in reply gladly give you all the relevant information I possess, though I regret to say that, in the way of positive information, the amount to be given is not large, the fact being that the existing state of knowledge on the subjects in question is very incomplete. . . . . . . In dealing with the force impressed on bodies by the water flowing past them, I hope I shall not appear too dogmatic or self-confident when I state that experiment and improved theory alike show that the resistance of such bodies as formulated in the ordinary text-books whether mathematical or practical, or on hydraulics, are almost invariably founded on erroneous hypotheses, and are incorrect. To begin with, it is usually stated that the resistance to a plane moving at right angles to itself through an inelastic fluid is equal to the weight of a column of the fluid having a sectional area equal to the area of the plane, and a length equal to the height due to the velocity; and it is on this assumption that Molesworth's coefficient . . . . . is founded. Beaufoy's experiments long ago proved conclusively that the real resistance exceeds them in the rates of 1.10 or 1.12 to 0.976.

"Again, the proposition that . . . . . the resistance to a cylinder moving at right angles to its axis, is half the pressure there would be on the diametric plane, calculated by the above erroneous hypothesis, is also an entirely mistaken proposition.

"I am able, from experiments of my own, to give an approximate measure of the resistance experienced by the cylinder, because in an investigation of the pressure log, as it is called, which I carried out with the Admiralty experimental apparatus, I ascertained the normal pressure on every part of the circumference of a cylinder thus moving.

"On analysing this series of pressures, it proved that their longitudinal component, that is the component in the line of motion, or rather the integral of thin longitudinal components, was just equal to the weight of a column of water, the base or sectional area of which is the diametric plane of the cylinder, and the height of which is the height due to the velocity. This component is not quite the whole resistance, for there must be added the component due to the circumferential drag of the water acting by surface friction. It is not easy to put a correct value on this addition, because the speed of the flow of the water past the surface of the cylinder is greatly modified by proper stream-line motion, and by induced eddies. I believe, however, that it is small in amount, and unless the surface become very foul it may be neglected.

"Taking the resistance in pounds, P; and the speed in knots, V; and the area of the diametric plane in square feet, A,

"The resistance of a plane moving at right angles to itself is about

$$P = 3.2\ V^2A \text{ in fresh water, and } 3.27\ V^2A \text{ in salt water.}$$

"The resistance of a cylinder moving at right angles to its axis is about

$$P = 2.78\ V^2A \text{ in fresh water, and } 2.85\ V^2A \text{ in salt water.}$$

"As regards the resistance of a sphere, Beaufoy's experiments give the result conclusively; I mean there seems no reason to mistrust their correctness. According to them we have as the resistance of a sphere

$$P = 1.0\ V^2A \text{ in fresh water, and } 1.03\ V^2A \text{ in salt water.}$$

"I should add that in all these cases it appears that the power of the velocity to which the resistance is proportional is rather under 2, in fact, about 1.87 to 1.95."

As the sphere is the best form of case for a buoyant body used in submarine mining, the most important formula of those given above is the last; and as the index of V is slightly less than 2, the constant may be reduced to unity, and the formula for the resistance P (lb.) of a sphere whose diametric area is A (sq. ft.) moored in salt water having a velocity V (knots per hour) becomes:

$$P = V^2\ A.$$

Difficulties engendered by Rise and Fall of Tide.—It will be noted in the example illustrated by the last diagram (Fig. 26) that when the rise and fall of tide approaches or exceeds the draught of the vessels which are to be acted upon by the mines, it is impossible to fulfil all the conditions required. Either the circuit-closing buoys (or the self-buoyant mines if these be used) must be so moored that they float on the surface at dead low water, or the vessels will be enabled to swim over the mines at high water. One way of meeting this difficulty is to moor the contact mines which are nearest to the mouth of a harbour in which the rise and fall of tide is considerable, at such a submersion that they will act during the lower half of the tidal level; and to moor the mines further up the harbour so that they will act during the upper half of the tidal level. When this plan is followed it is necessary to use a larger number of contact mines than would be required in a harbour with a moderate rise and fall of tide, and in order to obviate this difficulty a number of attempts have been made to design a trustworthy plan whereby the mines shall rise and fall automatically with the rise and fall of tide. The most successful arrangement is the following:

Rise and Fall Mines designed by Major R. M. Ruck, R.E.—In Figs. 27 and 28, A represents the floating body or mine, B is a counterpoise also possessing flotation, C is a chain graduated in size and weight, P is a pulley, S is a mooring, and D is a mooring rope or chain.

B may be a metal case either open at the bottom or closed by a waterproof diaphragm. B may also be a compressible waterproof closed bag, suitably weighted.

C is a chain made of links varying in weight; the larger links being furthest from B. A chain passing through the pulley is not indispensable; a weighted rope may be used with the heavier end resting on the bottom, A and B being connected in this case by a light wire rope passing round the pulley. Fig. 27 represents the system at low water, and Fig. 28 at high tide.

In shallow water two pulleys are required as shown in Fig. 29, and in waters where the depth is considerable as compared with the rise and fall of tide, the modifications shown in Figs. 30, 31, 32, can be adopted. When an electric cable is attached to the mine, it can be led away as shown in Fig. 34. In order to prevent any twisting action the mooring

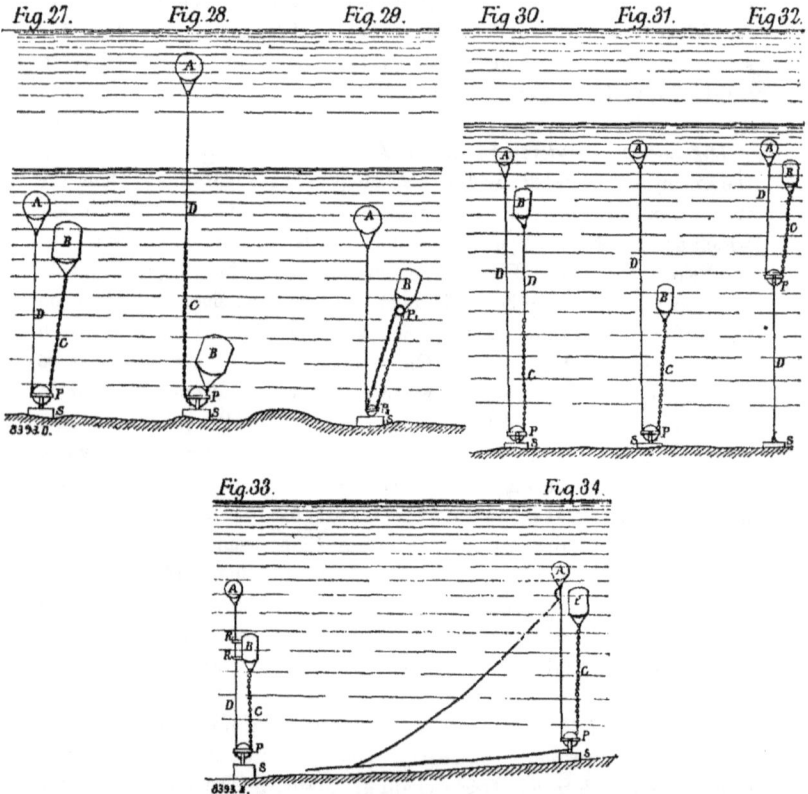

Fig. 27. Fig. 28. Fig. 29.

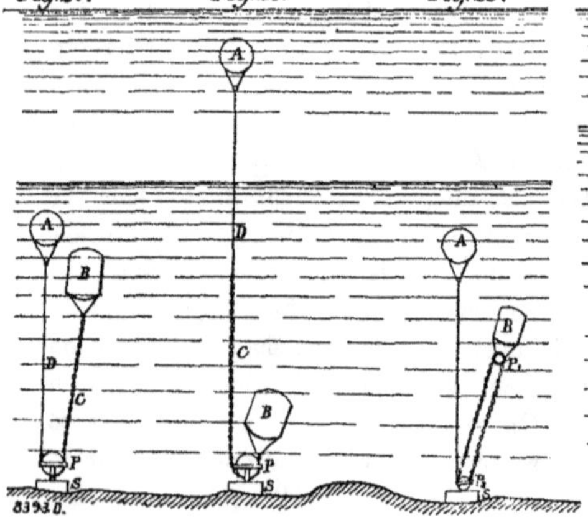

Fig. 30. Fig. 31. Fig. 32.

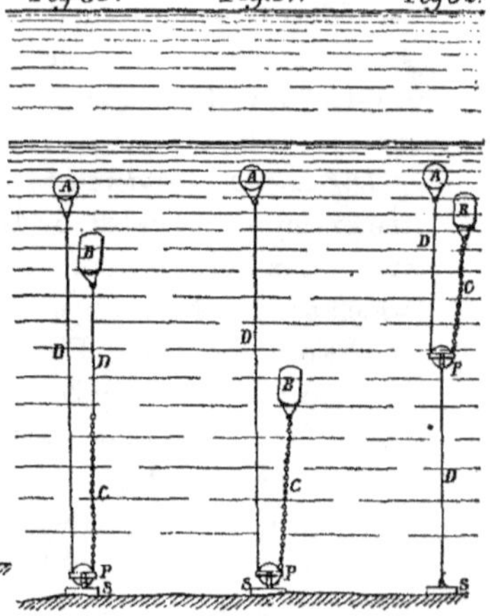

Fig. 33. Fig. 34.

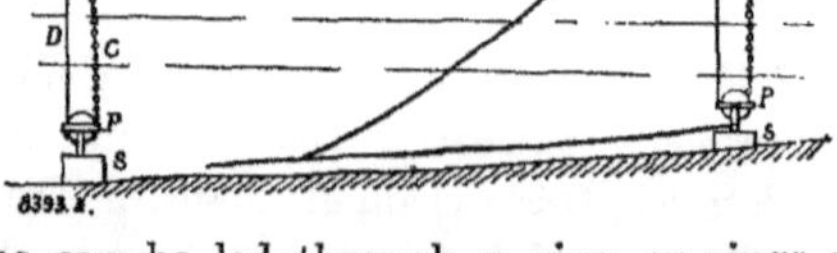

rope of the mine can be led through a ring or rings attached to the counterpoise as shown in Fig. 33. Several other modifications have been described by the inventor, but the simpler forms now illustrated are probably of the most practical value. In order to assure success with this system, care must be taken in the manufacture of the gear and in laying the mines.

The action is as follows: Commencing, say, at low water, as the tide rises the increased water pressure on the counterpoise reduces its flotation by compressing the air contained in it. The equilibrium of the system is overthrown; B sinks, A rises, until equilibrium is restored by some of the heavy chain passing round the pulley or by some of the weighted rope falling and resting upon the ground. Another rise of tide occurs, and the action is repeated. A fall of tide produces an action in the opposite direction. By these means the mine is kept within narrow limits at a constant depth below the surface. The mine and the counterpoise should be so arranged that they never touch one another. Should this occur in rough weather they damage one another.

The inherent difficulties of the problem have been ingeniously met in this solution by Major Ruck, but the gear is necessarily heavier and more difficult to work than the usual arrangements, and the mooring must be twice as heavy as usual. It is never necessary to employ automatic arrangements for mines to catch large vessels where the rise and fall of tide does not exceed three fathoms, and when this is exceeded, the tidal currents are so powerful that it becomes a matter for consideration whether to adopt automatic rise and fall mines, or to employ a larger number of ordinary mines at different levels, or to use observation mines instead of mines that are fired by contact. The local peculiarities of each harbour must be thoroughly examined before deciding on such a matter.

Spacing of Electro-Contact Mines.—The next question to which an answer is required, refers to the distance that should separate electro-contact mines, and the following considerations should be borne in mind before a reply is given.

1. The mines should be spaced at such intervals that there is no danger of their fouling one another when eddies set them in opposite directions.

It is highly improbable that these eddies will ever simultaneously act in such a manner as to tilt mines on single moorings so that the angle a with the vertical (see Fig. 26) will attain 18 deg., the mines being tilted in opposite directions. But, assuming that this extreme case is possible and that the length of the mooring line is also extreme, say 20 fathoms; then the mines cannot foul each other if the sinkers be separated by a distance of 12½ fathoms, or 75 ft. Other considerations to follow will show that electro-contact mines must be spaced at greater intervals than 75 ft.; but the above is important, because the mine intervals, whatever they may be fixed at, for other considerations should be increased by an amount not exceeding $\frac{12\frac{1}{2}}{20} = 0.625$ of the vertical length of the mooring line, if the site be one in which strong sworls or eddies of water occur at any time of tide.

2. The mines should be spaced at such intervals that the explosion of one mine shall not damage any of the neighbouring mines. This consideration shows the advantage of employing the smallest effective charges in electro-contact and other buoyant mines.

If a charge of 72 lb. of blasting gelatine be employed, as before suggested in foot-note p. 59, at what distance will the spherical circuit-closer jacket be safe from damage?

As already shown, a 3-ft. sphere, made of $\frac{1}{4}$-in. steel, should withstand a collapsing pressure of 1400 lb. on the square inch; and it has been proposed in these papers to make all spherical cases for buoyant mines, circuit-closers, &c., of the same strength to withstand such pressures.

This value for P is obtained by the subaqueous explosion of 72 lb. of blasting gelatine at a distance of about 50 ft. horizontally from the charge, as calculated by the author's formula already given.

Hence to meet consideration (2) the electro-contact mines under discussion must be spaced apart at 50 ft. And if the depth of water be such that mooring lines 10 fathoms long are required, 0.6×60 ft. = 36 ft. must be added to meet consideration (1). The total spacing must therefore be at least 50 ft. + 36 = 86 ft. for the mines under discussion.

3. The mines must also be so spaced that when one is exploded it shall not cause the neighbouring mines to signal as if struck by a vessel, for this would cause them to explode also, and thus the whole of the electro-contact mines in one group might be exploded simultaneously, when only one should explode. Electrical arrangements can, however, be made on shore at the firing station, which will prevent this undesirable result, and the matter need not be further discussed here except to state that even in the absence of such electrical safeguard, there is no practical difficulty in so adjusting the circuit-closing arrrangements of the mine, that no signal of a neighbouring mine shall be caused by a mine's explosion when the mines are spaced at intervals of 100 ft., the minimum spacing for electro-contact mines should therefore be 136 ft., when the charges are limited to the amount stated. As this distance will also meet considerations (1) and (2) in the example taken, it may be accepted for depths up to 10 fathoms. Beyond this depth the spaces between the mines should be slightly increased, say to 150 ft.

4. There yet remains another matter of some practical importance that bears upon this question, viz., the length and handiness of the steamer employed for laying the mines. After a long experience I am

convinced that the length of such a craft ought not to exceed 70 ft., and she should be a very handy ship. If the mooring steamer be 90 ft. or 100 ft. long, she is liable to foul and drag the mines already laid, when carrying out the mooring operations. This difficulty is, however, avoided by what is termed the dormant system, which will now be described.

Dormant Electro-Contact Mines.—It will frequently occur that mines must be laid in channels which cannot be closed to commerce or to the frequent passage of friendly war vessels, and yet it may be very desirable to employ electro-contact mines rather than observation mines for which there may be no suitable sites for observing stations in the vicinity of the said channels. Under the above conditions the ordinary electro-contact mines are evidently inapplicable, as they would be cut away and destroyed by the screws of passing steamers, and might even be accidentally exploded, if the detonating fuzes were subjected to severe blows. Buoyant mines can then with advantage be moored in such a manner that they are held down to the bottom until they

Fig. 35.

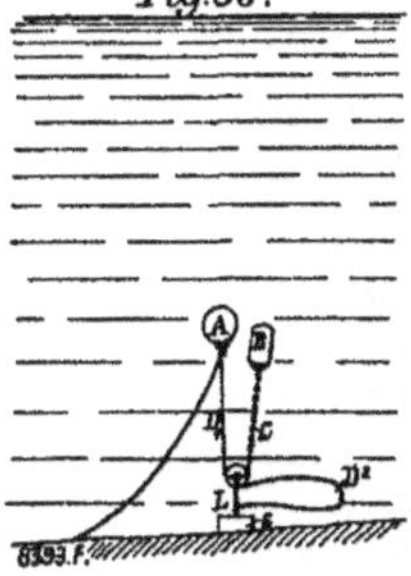

Fig. 36.

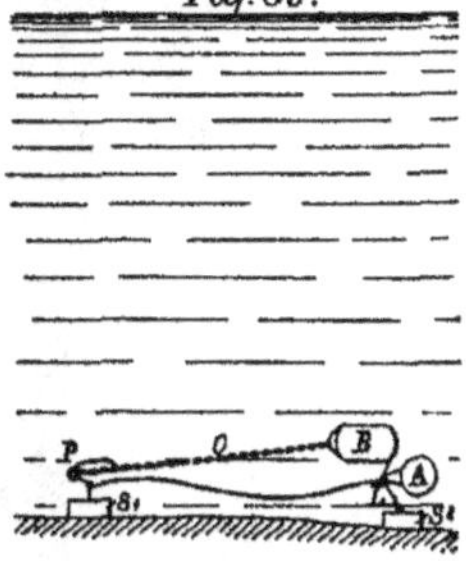

are required to rise into the positions required to prevent the passage of hostile vessels. They are then called dormant mines. The plan can be employed in connection with mines on a single mooring and sinker, or with mines on a double mooring or bridle and two sinkers, or with mines arranged on Major Ruck's system, as shown on Figs. 35 and 36, in which L L are explosive links fired by a suitable electric current in such a manner that the mines are not exploded when the links are exploded. The other letters on these figures refer to the articles similarly lettered and already described in the rise and fall system of electro-contact mines. When but one sinker is used, precautions must be taken to prevent the slack portion of the mooring line from fouling the rest of the gear, or the mine, &c., will not rise into the proper place when the explosive link is fired. One method that suggests itself is to coil the slack of the mooring line on the top of the sinker, or around it, and to tie the coil with weak stoppers, which the buoyancy of the mine, &c., can break as soon as the link is fired.

Another plan that suggests itself is to coil the slack on a wooden drum, one end of which is secured to the sinker by a short piece of rope, and the other by an explosive link. On the latter being fired the drum would tilt into a vertical position, and the mooring line be released.

When a mine is moored on a bridle between two sinkers, an eye can be formed two or three fathoms down the wire rope mooring line, and the mine be secured to it by an explosive link. This is a simple arrangement, and the dormant system therefore seems to be well adapted for mines moored in this manner.

The explosive link was first suggested by the author when carrying

Fig. 37.

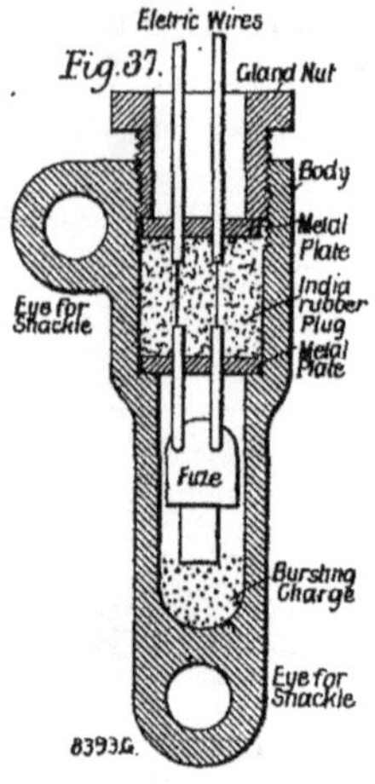

out the experiments against H.M.S. Oberon, and was afterwards embodied in his mechanical system of submarine mines. The links consist of short iron or metal tubes containing a small bursting charge and an electric fuze; also suitable water-tight entrances for the electric wires and metal eyes cast on the body of the tube to take the necessary shackles or wire lashings. The interior should be turned out where the india-rubber plug rests, to insure a water-tight joint. A bursting charge of 1 drachm of rifle powder is sufficient, the body being $\frac{1}{4}$ in. thick if made of cast iron, and $\frac{1}{8}$ in. thick if made of brass. The electrical arrangements in the mine and circuit-closer are such that the explosive link for any particular mine can be fired when it is desired to do so without firing the mine itself, although a single cable is used. The electrical details for electro-contact and other mines must be explained hereafter.

Manufacture of Cases for Electro-Contact Mines.—The size and thickness and shape of the cases having been settled, after a decision has been arrived at concerning the nature of the explosive to be employed, no important difficulty is likely to arise in their manufacture. Siemens Landore steel is usually employed, the hemispheres being pressed into shape when hot by stamps worked by hydraulic rams. It is usual to secure eyes by means of palms rivetted to the case afterwards, but these palms are a source of trouble, the cases being apt to leak at the palms, after any rough handling to which they may be unavoidably subjected. It is easy to avoid the use of such eyes, providing instead, two wire rope rings in which eyes are made, and connecting these rings by wire rope bracing as shown on Fig. 38.

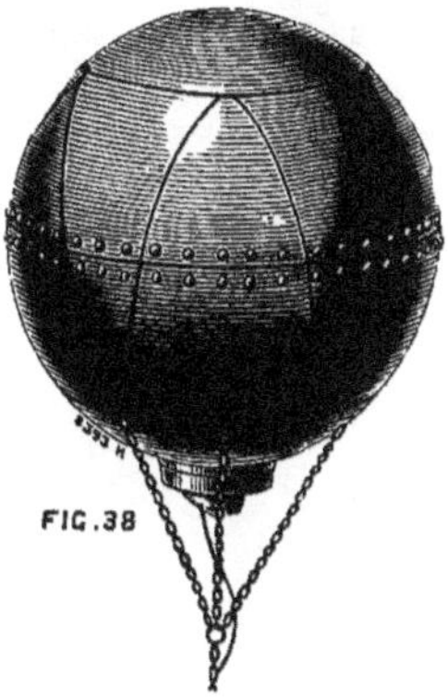

FIG. 38

Another point of weakness in buoyant cases for submarine mining is the existence of rivetted joints; for although it is easy enough to make the joints water-tight so as to withstand ordinary rough usage, it must be borne in mind that such cases have to remain water-tight after receiving the severe blows occasioned by mines exploding in their vicinity. Rivetted joints should therefore be avoided if possible, because a small leak soon causes the mine case to fill and to sink.

Tinned joints would be absolutely water-tight, should be as strong as rivetted joints, and would be lighter. Rivetted joints are apt to leak because that portion of the case being more rigid than the remainder, any indentation caused by a neighbouring explosion is apt to pull open the joint by bending the steel plate inwards away from the joint, and outwards at the joint or caulking, the line of rivets being the fulcrum.

With a tinned joint, the lips would be the strongest portion of the connection.

As regards the mine cases in Division 2, Class A, viz., those not under control and coming under the generally accepted name of "mechanical mines" (although some are electrical and others chemical), they are generally made as cheaply as possible so that their numbers may compensate for inferior individual efficiency as compared with the mines under control. Moreover, inasmuch as they must be spaced at such intervals that the explosion of one shall not cause the neighbouring mines to act, the cases can be much weaker than those for electro-contact mines where electrical arrangements can be introduced to prevent such self-destruction. As "mechanical mines" will be examined in a separate chapter, no more need be said at present about them.

The manufacture of the cases for large buoyant mines will be treated in the next chapter.

CHAPTER VI.

LARGE MINES.

THE cases required for Class B mines, viz., those that contain large charges and act at a distance from the target, will now be considered.

All these mines are necessarily in Division 1; *i.e.*, are under control, and are therefore electrical mines. But they can be conveniently divided into two sub-divisions: (1) Ground mines, which lie on the bottom; (2) buoyant mines.

Ground mines are invariably used in preference to the latter when the depth of water is not excessive.

French System.—As before stated, the French appear to use ground mines up to depths of 80 ft., the charges being:

TABLE XXII.

550 lb.	gun-cotton,	or 2200 lb.	gunpowder,	26 ft.	to 36 ft.
660	,, ,,	3300	,, ,,		up to 50 ,,
880	,, ,,	4400	,, ,,		,, 60 ,,
1100	,, ,,				,, 67 ,,
1320	,, ,,				,, 73 ,,
1540	,, ,,				,, 80 ,,

English System.—The English custom has remained practically unaltered since 1873, when Lieutenant-Colonel R. H. Stotherd's book, "Notes on Submarine Mines," was published in England and reproduced soon afterwards by an enterprising publisher in New York, U.S.A. The charges for ground mines therein mentioned for the following depths are:

TABLE XXIII.

250 lb.	gun-cotton,	20 ft.	to 35 ft.
500	,, ,,	35 ,,	60 ,,

When the water is over 60 ft. in depth, 500 lb. buoyant mines are used, and are moored about 48 ft. from the surface. These charges and limits were not altered after the completion of the series of experiments against the Oberon.

American System.—It is believed that the limit of size of mine charges in the American service has been fixed at 400 lb., and that American adepts prefer to employ a number of moderate charges rather than a few large mines. But the arguments (already advanced in the paper on contact mines in this series) advocating the employment of the minimum effective charges when the mines are buoyant, do not apply when the mines rest on the bottom; for there is then no difficulty whatever in making the cases very strong to resist neighbouring explosions, and when the mines are fired by observers at a distance the large mines are more likely to act in the desired manner, on account of larger area of effect.

Small charge observation mines cannot therefore be recommended, except in comparatively shallow waters, which are situated near to the observing stations—and, as a rule, these waters can be mined more effectively by electro-contact mines—or by small ground mines with detached circuit-closers. A 250-lb. mine charged with gun-cotton possesses an effective striking power up to distances of 20 ft. from its centre; consequently its effective circle for observation firing must always be less than 40 ft. in diameter, which maximum would only be attained by a ground mine when the vessel attacked is wall-sided, flat-bottomed, and drawing nearly the full depth of the mined water. These considerations must make it evident that ground mines with charges not exceeding 250 lb. gun-cotton are only adapted for observation firing at very close quarters.

Ground Mines Fitted with Detached Circuit-Closers.—Small ground mines can, however, be usefully employed when they are fired by means of a detached circuit-closer moored above them. For instance, the 250-lb. gun-cotton ground mine would act well in water up to 45 ft. against vessels drawing 25 ft. and over, the detached circuit-closer being 20 ft. above the bottom; and if blasting gelatine were used in the same cases they would be effective up to depths of 53 ft. Similarly, a ground mine charged with 500 lb. of gun-cotton, if fired by a detached circuit-closer moored 38 ft. above it, that being the striking distance of such a mine against an ironclad, would be effective against vessels of 25 ft. draught and over in waters up to 63 ft., and if blasting gelatine were used the mines would act up to depths of 78 ft. But the employment of detached circuit-closers in connection with mines containing large charges has to a great extent gone out of fashion, because such mines are generally placed in those deeper portions of a harbour which cannot conveniently be obstructed by buoyant bodies which may foul, or be themselves damaged by passing vessels that are not foes. This objection can, however, be readily met by the dormant system of mooring mines already referred to, and the defence then obtained appears to be formidable, and likely to be resorted to, especially with ground mines and detached circuit-closers.

Ground Mines Fired by Observation.—Ground mines which are fired by observation, *i.e.*, by an observer (or by two observers) from a distance, when the vessel attacked is seen to be near enough to the mine, should evidently have a horizontal circle of effect sufficiently large for the proper working of the observation instruments, and this will depend upon the distance separating the mines and the instruments and upon the accuracy of the latter. Assume that the mines are about one sea mile from the instruments and that an effective horizontal circle of 30 ft. radius is desired. Also, assume that the depth of water is 60 ft., that the draught of an ironclad is 27 ft., and that the shape of her side is somewhat as shown in Fig. 39. A geometrical construction will then show that the actual distance between the mine and the nearest point of the vessel is 52 ft. If the mines were situated only half

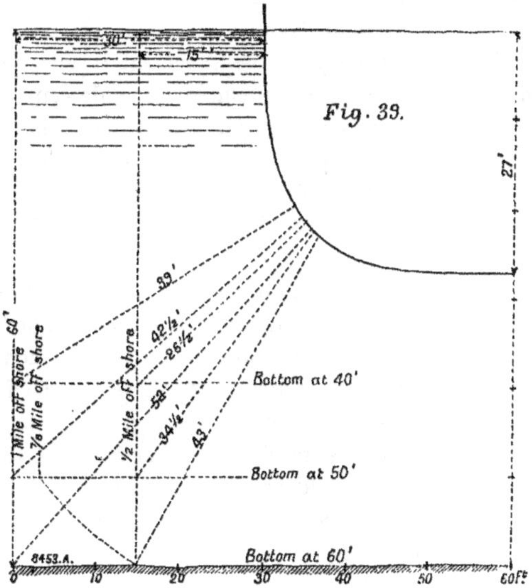

Fig. 39.

a mile from the observing instruments, equal accuracy would be obtained with mines having a horizontal circle of effect of half the former dimensions, and the striking distance is then reduced to 43 ft., and the charges can be reduced in nearly the same proportion, or about four-fifths of those for 52 ft. By increasing the depth shown on the sketch it will be found by measurement that the following Table gives the striking distances required for effective horizontal circles of 30 ft. and 15 ft. radius respectively when the vessel draws 27 ft. of water.

Similarly, if the depth of water be 50 ft., instead of 60 ft., the striking distance for mines one mile off should be about 42½ ft., and for mines at half a mile the striking distance should be about 34½ ft., and the charges may be regulated accordingly.

TABLE XXIV.

Depth in Feet.	Striking Distances in Feet. Giving a Circle 30 ft. Radius.	Striking Distances in Feet. Giving a Circle. 15 ft. Radius.
40	39	26½
50	42½	34½
60	52	43
70	60	51
80	68	60
90	76	69
100	85	78

Again, a mine which is effective in 60 ft. of water at half a mile off shore would be equally effective in 50 ft. of water at seven-eighths of a mile off shore. And so the changes can be rung.

In the defence of some harbours by submarine mines a considerable economy can be made by keeping these matters in view, but it is usual to sacrifice such economy and to employ one description of ground mine for all situations, thus avoiding numerous patterns.

From a Table given in a previous chapter it will be found that the minimum charges for an effective strike of 52 ft., should be 484 lb. blasting gelatine, or 558 lb. gelatine dynamite, or 687 lb. gun-cotton or dynamite No. 1.

It is not prudent to rely upon observation firing at greater distances than one sea mile, the smoke of an engagement, or fog, or thick weather having to be reckoned with in this class of mine. If then we limit the employment of ground mines to water not exceeding 60 ft. in depth, a case which will hold 484 lb. of blasting gelatine would appear to meet all requirements. But this explosive is 45 per cent. heavier than gun-cotton, and the same case would therefore hold only 334 lb. gun-cotton, the striking distance of which against an ironclad is only 26 ft., barely sufficient for an observation ground mine half a mile off shore in 40 ft. of water, and insufficient for greater depths or distances. On the other hand a case to contain 500 lb. of gun-cotton will hold 725 lb. of blasting gelatine, the former having a striking distance of 38 ft. and the latter of 77 ft. when acting against a modern war vessel. If, therefore, it be intended to use either or both of these explosives in a time of emergency, it would appear that a convenient arrangement would be obtained by having two patterns for ground mines. Adding 20 per cent. to the above figures, so as to be on the safe side, the one should be made large enough to hold either 600 lb. of gun-cotton or 870 lb. of blasting gelatine, and the other to hold 600 lb. of blasting gelatine or 414 lb. of gun-cotton. As the slab gun-cotton presents flat surfaces to the curve of the cylinder, each case would hold a little more of the blasting gelatine than denoted by the above figures, and it would be nearer the truth to put round figures as shown on the Table to follow.

(N.B.—Although blasting gelatine is superior to dynamite and gelatine dynamite, it may occur that these explosives will be used in a time of emergency, and the cases would contain about the same weights of them as of blasting gelatine.)

This Table gives a useful *technica memoria* for the charges of large mines, viz., to use 10 lb. of blasting gelatine per foot of strike required. Also that the same case will hold two-thirds as much gun-cotton by weight possessing half the strike in feet. By a diagram similar to the one given it can be shown that a mine with a strike of 90 ft. can be used in 105 ft. of water, at one mile from an observing station, an effective horizontal circle of 30 ft. radius against vessels of 27 ft. draught being obtained.

The employment of blasting gelatine in ground mines is evidently extremely advantageous when the waters are deep, and it would appear that the rule in the English service limiting the employment of ground mines to depths of 60 ft. or thereabouts would be improved if a larger limit (say, 90 ft. or 100 ft.) were fixed upon, ground mines being simpler than buoyant mines, and less liable to derangement.

Best Shape for Ground Mines.—As a ground mine may be heavy the cheapest form of case is a cylinder, and Staffordshire plate may be used in its manufacture, the required strength being obtained by making the skin of a good thickness, and the diameter of the case as small as is practicable. In the English service the ground mine most usually employed holds 500 lb. of gun-cotton. It is described and illustrated in Colonel (now General) R. H. Stotherd's book as a cylindrical iron case ¼ in. thick, 34 in. side, 30 in. in diameter, and its ends are dished out with a radius of 30 in. No important alterations have been made in this mine case, but it has been strengthened internally by a lining made of Portland cement, experience having shown that the case is weak when subjected to countermining. New cases should, however, possess the requisite strength from their thickness, shape, size, and method of manufacture. A cylinder having an internal radius of 1 ft. 1¾ in. is convenient for loading with the English slabs of compressed gun-cotton, and the interior should be a true and smooth cylinder, a longitudinal butt joint being employed, with the covering strip outside. The ends should not be dished outwards, but be plane surfaces, or dished slightly inwards, and the end joints should project, thus affording an opportunity for hydraulic rivetting, and making a very strong job.

With an internal diameter of 2 ft. 3½ in., each layer of compressed gun-cotton 1¾ in. thick will contain 14 slabs of the explosive in its

English form, two of them being sawn across diagonally, and the centre slabs being sawn as required for the introduction of the priming charge and firing apparatus (see Fig. 40). This gives about 35 lb. per layer, 17 layers for the 600 lb., and a total length of 2 ft. 6 in.

For the smaller cylinder, as slab gun-cotton stows well in a circle 1 ft. 9 in. in diameter (see Fig. 41), that dimension will be chosen for the internal diameter of the case. Each layer of gun-cotton will then contain $8\frac{1}{4}$ slabs, $20\frac{1}{2}$ lb. per layer $1\frac{3}{4}$ in. thick, $19\frac{1}{2}$ layers for 400 lb., and a total length of 2 ft. 10 in.

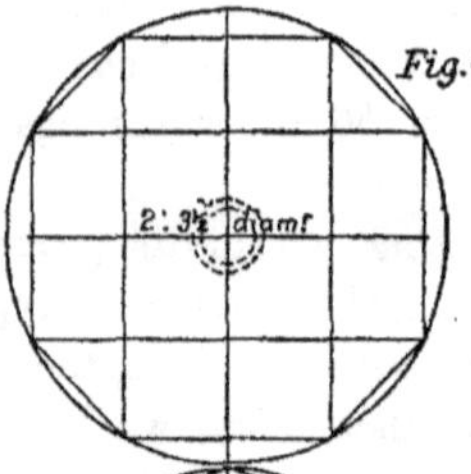

Fig. 40.

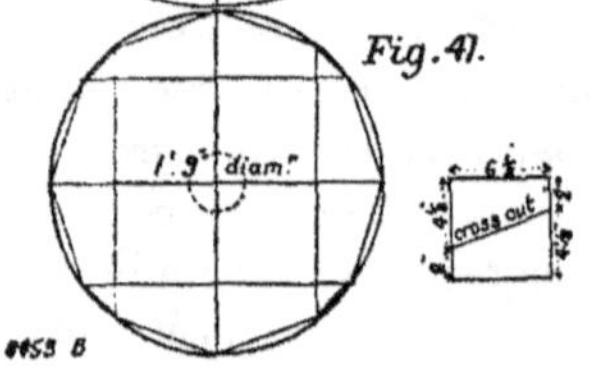

Fig. 41.

Case I., therefore, is 27.5 in. in diameter and 30 in. long inside, and Case II. is 21 in. in diameter and 34 in. long inside.

In loading these cases with slab gun-cotton the last three layers must be composed of quarter slabs. It is better to do this than to use a larger loading hole, as the size and weight of the door or "mouth-piece" should be made as small as possible, because weight is of importance in the buoyant mines, and the apparatus for the ground and buoyant observation mines should be interchangeable, thus reducing the number of patterns.

What thickness of iron should be used in these cases to make them as strong as a 3-ft. sphere of $\frac{1}{4}$-in. steel? Now the strength of a sphere is twice that of a tube of equal diameter (Rankine), and the strength of the sphere varies as the square of thickness of shell, and inversely as the square of the diameter, or as $\frac{t^2}{d^2}$. Also the strength of the sides of a tube varies as $\frac{t}{l\,d}$ (l being the length).

Consequently:

$$\frac{\frac{1}{2}t^2}{30\times 27.5}=\frac{(\frac{1}{4})^2}{(36)^2} \text{ and } t=0.29 \text{ in.}$$

Similarly for the smaller case:

$$\frac{\frac{1}{2}t^2}{21\times 34}=\frac{(\frac{1}{4})^2}{(36)^2} \text{ and } t=0.31 \text{ in.}$$

This assumes that the cases are made of the best and toughest steel. But it is proposed to make them of wrought iron, the strength of which is but little more than half that of steel. Hence for the large cylinder:

$$t^2=2\,(0.29)^2, \text{ and } t=0.41 \text{ in. of iron.}$$

And for the smaller cylinder:

$$t^2=2\,(0.31)^2, \text{ and } t=0.44 \text{ in. of iron.}$$

The flat ends of each cylinder should be made slightly thicker, say $\frac{1}{2}$ in. thick. There can be no objection on the score of weight in making the whole of each case of $\frac{1}{2}$-in. iron, and it would be advantageous to do so.

The projecting joints between the ends and sides will make the cylinders about 5 in. or 6 in. longer, and the weights, &c., are shown in the following Table, which also gives other useful information:

TABLE XXV.—CYLINDRICAL CASES FOR GROUND MINES.

Description.	Large Case.	Small Case.
Material: $\frac{1}{2}$-in. Staffordshire iron plate.		
Length of cylinder, inside	30 in.	34 in.
,, ,, case over all	36 ,,	40 ,,
Diameter, inside	27½ ,,	21 ,,
Side surface of case, including strip for butt joint ...	23 sq. ft.	20 sq. ft.
End surfaces with turnover...	11.2 ,,	7.1 ,,
Weight of case empty in air	684 lb.	540 lb.
,, ,, salt water displaced	710 ,,	480 ,,
,, ,, charge blasting gelatine (dynamite or gelatine dynamite)	900 ,,	600 ,,
Weight of charge gun-cotton slabs...	600 ,,	400 ,,
Effective Strike to a Man-of-War.		
When loaded with blast gelatine	90 ft.	60 ft.
,, ,, ,, gelatine dynamite	78 ,,	52 ,,
,, ,, ,, dynamite	67 ,,	45 ,,
,, ,, ,, gun-cotton slabs	45 ,,	30 ,,

The loading hole should be 6½ in. in diameter if English pattern gun-cotton slabs are to be used. Lugs for attachment chains should not be rivetted to the case by palms, as it is difficult to keep them water-tight. If the method of constructing the cylinders now proposed be followed, it would be easy to weld projections or ears on the turnover of flat end pieces, and to provide each ear with a ring.

In order to prevent a cylindrical case from rolling about on the bottom when laid, and thus not only getting out of position, but in all probability damaging the electric cable at its point of attachment, it is necessary to fix some sort of cradle, or lash two short spars to the case, one on either side of it, thus forming a rough but efficient cradle.

The foregoing descriptions have been given so much in detail, that it will be unnecessary to show any drawings of these mine cases.

Large Buoyant Mines.—When very deep waters have to be mined with large charges, it is necessary to employ buoyant cases, and most of the observations which have already been made on small contact mines are equally applicable to large buoyant mines. It will be desirable in the first place, therefore, to settle upon an effective minimum for the charge to be employed. The explosive employed should certainly be the most powerful obtainable.

Submersion.—Assuming that the mines are buoyed up to a submersion of 40 ft. (see Fig. 42) at low-water springs, and that the rise and

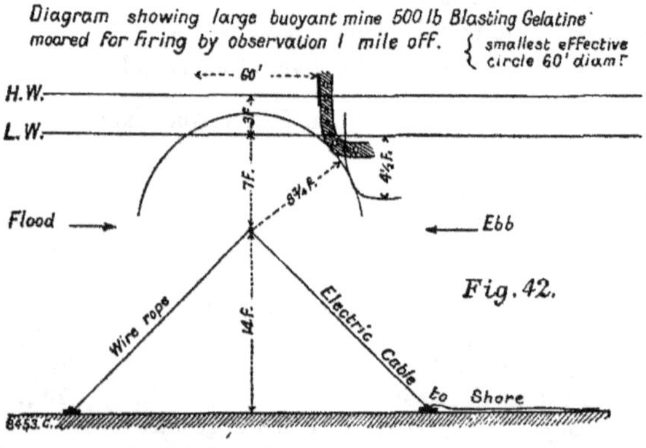

Fig. 42.

fall of tide amounts to 20 ft., the mines would sometimes be submerged 60 ft., and should therefore possess an effective striking distance of 52 ft. if moored a mile from the observing station, and of 43 ft. if moored half a mile off. If the rise and fall of tide were less than 20 ft. the submersion at low water may be increased by the difference. Thus if it were 12 ft. rise and fall, the submersion may be 48 ft. at low water. The maximum striking distance of 52 ft. would thus remain unaltered.

Charge required.—A charge of 485 lb. of blasting gelatine possesses the requisite power, as also does a charge of 687 lb. of gun-cotton, but the former is much to be preferred, because a smaller buoyant body is required to support it, and this is important. Moreover, the matter of cost is considerably in favour of the more powerful explosive; 500 lb. of blasting gelatine will therefore be selected as the normal charge for a large buoyant mine.

An objection has been urged in these pages against the use of an air space round the small charges for contact torpedoes, but the same arguments do not apply to large charges acting at a much greater distance when the blow delivered is of a racking rather than of a punching character. The charge may therefore be held in the buoyant case, and the shape of this case should be spherical for the reasons already given when discussing contact mines. The buoyancy is then a maximum for a given weight of case, and the resistance offered to the current is a minimum. The double mooring previously advocated for contact mines is still more essential for observation mines, accuracy of position being so important. Proceeding as in the last example for contact mines to calculate the necessary size of case we find that

$$14\,B - 14\,P - 7\,P^1 = 0,$$

where B is the available buoyancy in pounds of the mine when loaded and moored, P is the side pressure in pounds due to the current acting on the mine, and P^1 is the side pressure on the wire rope and cable, span being 90 deg., and the mine 14 fathoms above the bottom.

Now $$P = 1.03\,V^2 A,$$

A being the diametric area of the sphere,

And $$P^1 = 2.85\,V^2 A^1,$$

A^1 being the area of the diametric planes of the cable and wire rope resolved into the vertical.

$$\therefore A^1 = 20 \times 6 \times \frac{8}{10 \times 12} \cos 45 + 20 \times 6 \times \frac{2}{3 \times 12} \cos 45 = 11 \text{ sq. ft.}$$

Substituting

$$B = 1.03\,V^2 A + \tfrac{1}{2}(2.85\,V^2 \times 11)$$
$$= V^2 (1.03\,A + 15.67)$$

When $V = 2$ knots per hour,
$B = 4.12\,A + 62.68.$

When $V = 3$ knots per hour,
$B = 9.27\,A + 141.13.$

When $V = 4$ knots per hour,
$B = 16.48\,A + 250.72.$

When $V = 5$ knots per hour,
$B = 25.75\,A + 391.75.$

But the mine has to support 500 lb. of blasting gelatine, 57 lb. of cable, and 68 lb of wire rope, or 625 lb. in all, in addition to its own weight. Hence the buoyancy when empty

$= 4.12\,A + 688$ for 2 knots.
$= 9.27\,A + 766$,, 3 ,,
$= 16.48\,A + 876$,, 4 ,,
$= 25.75\,A + 1017$,, 5 ,,

Size of Cases for Large Buoyant Mines.—Referring to Table XX. for steel spheres with a thickness of skin $= \frac{D}{144}$, it will be found that these equations are satisfied when D, the diameter of the case, is 3.2 ft. for a 2-knot tide, 3.4 ft. for 3 knots, 3.6 ft. for 4 knots, and 3.85 ft. for 5 knots.

The tidal currents most usually met with on harbour mine fields do not exceed 2 knots per hour; a 42-in spherical case of $\frac{7}{24}$-in. steel, which is effective in waters up to $3\frac{1}{2}$ knots per hour, will therefore answer well in most situations. When the limit of $3\frac{1}{2}$ knots is exceeded the case employed should be a 4-ft. sphere of $\frac{1}{3}$-in. steel, which is effective in currents up to $5\frac{1}{2}$ knots, the angle of span and the method of mooring the mines hereinbefore suggested being followed.

Inasmuch as these mines must be loaded with blasting gelatine, and that it would be difficult to insert the apparatus for firing the charge into the bottom of the mine after such a charge has been inserted, it will be advisable to provide a loading hole in the top of the case, in addition to the hole usually made in the bottom for the apparatus. The charge can then be inserted after the apparatus, and can be packed closely round its envelope; also, the top surface of the explosive can be covered with a light deck of thin wood, kept in position by suitable battens and struts.

The writer recommends that the mines be attached to their mooring lines by an encircling ring made of wire rope of smaller diameter than the case, such ring being provided with thimbles rove into it, and the ring kept in place by means of a second and similar ring and braces of wire rope between them—as was recommended for the spherical buoys employed in connection with contact mines. In this manner the employment of ears or palms rivetted to the case is avoided.

The following Table of the principal details of the two buoyant mine cases which it is proposed to adopt will be useful for reference:

TABLE XXVI.—SPHERICAL CASES FOR BUOYANT MINES.

Description.	Large Case.	Small Case.
Material: Landore steel, diameter = 144 thickness.		
Diameter of sphere inside	48 in.	42 in.
Surface of sphere	50.4 sq. ft.	37.7 sq. ft.
Weight of case empty in air	672 lb.	440 lb.
,, ,, salt water displaced	2144 ,,	1427 ,,
,, ,, charge (blasting gelatine preferred)	500 ,,	500 ,,
,, ,, apparatus...	30 ,,	30 ,,
,, ,, wire rope at 45 deg. per fathom deep ...	4.9 ,,	4.9 ,,
,, ,, cable at 45 deg. per fathom deep	4.1 ,,	4.1 ,,
Area of diametric plane	12.57 sq. ft.	9.6 sq. ft.

Effective Strike to a Man-of-War.

Description.	Large Case.	Small Case.
When loaded with blasting gelatine	54 ft.	54 ft.
,, ,, ,, gelatine dynamite	47 ,,	47 ,,
,, ,, ,, dynamite	38 ,,	38 ,,
,, ,, ,, gun-cotton slabs	38 ,,	38 ,,

The large mines recommended should therefore be spaced as follows :

TABLE XXVII.—SPACING FOR LARGE MINES.

	At or over
Buoyant, large and small sizes, 500 lb. blasting gelatine ...	456 ft.
,, ,, ,, ,, ,, ,, gun-cotton	321 ,,
Ground, ,, size, 900 lb. blasting gelatine	387 ,,
,, ,, ,, 600 ,, gun-cotton	184 ,,
,, small ,, 600 ,, blasting gelatine	228 ,,
,, ,, ,, 400 ,, gun-cotton	91 ,,

Before leaving the cases for submarine mines, a few words are desirable on the distance that should separate the large observation mines. Of the four considerations bearing upon this point with respect to contact mines, only one applies to the observation mines, viz., that they shall be spaced at such minimum intervals that the explosion of one mine cannot injure its neighbours. Assuming that the apparatus used in the mines is strong enough to stand any shock that will not injure the cases themselves (and care should be taken that this is so, a care that is too frequently neglected), it only remains to calculate the distances at which the charges proposed to be employed will damage the cases described. The buoyant cases, large and small, are designed to be equal in strength to a 3-ft. steel sphere $\frac{1}{4}$ in. thick, which will withstand a collapsing pressure of 1400 lb. on the square inch (see top of page 59). Now 500 lb. of blasting gelatine will produce this pressure in water at a distance of about 456 ft., *vide* formula, page 36, and 500 lb. of gun-cotton will do the same at 321 ft. Buoyant mines of the patterns recommended should consequently be moored at intervals not smaller than the above, according to the explosive employed in them. With regard to the ground mines, it can be shown by the same formulæ that the large iron case can resist a collapsing pressure of 2931 lb. on square inch, and the smaller case one of 3364 lb. Also, that 900 lb. of blasting gelatine will produce the former effect at 387 ft., and 600 lb. of gun-cotton at 184 ft. Also, that 600 lb. of blasting gelatine will produce the latter effect at 228 ft., and 400 lb. of gun-cotton at 91 ft.

Conclusion.—The various descriptions of mines under control and of buoys for circuit-closers have now been examined as to shape, size, weight, displacement, thickness, nature of material, resistance to currents, spacing, &c.

Messrs. Day, Summers, and Co., of the Northam Iron Works, Southampton, have undertaken to manufacture them, and any additional information can be obtained on application to this firm. The methods of mooring them have been roughly indicated, and will now be examined more in detail, the considerations and diagrams which should settle many matters connected therewith being fresh in the mind of the reader.

CHAPTER VII.

Mooring Gear for Submarine Mine.

Sinkers.—As already stated, ground mines when made of thick iron are sufficiently heavy to keep their position as laid, without providing sinkers for them, as was customary a few years ago, the development of countermining showing that it was a better arrangement to place weight in the case than in a sinker outside the case. In this way ground mines can be made much stronger than they used to be, and yet be handled and moored with equal facility. Mooring sinkers for ground mines are therefore no longer required, but mines which are buoyed up from the bottom (whether they be contact mines or mines fired by detached circuit-closers or large mines fired by observation) must evidently be anchored to the bottom in such a manner that there is no probability of their moving from the positions in which they are laid. Ordinary anchors have seldom, if ever, been employed for this work, because mines are generally moored by single moorings, for which purpose an anchor is unsuitable. When double moorings are used anchors may be employed, but it is more difficult to place them exactly in position than dead-weight sinkers, and even if the top fluke be turned down they are more likely to be fouled and dragged out of position afterwards. Moreover, on hard bottoms they are more liable to shift their positions under normal conditions than would sinkers of suitable weights. If, however, sinkers were not procurable in sufficient numbers on an emergency, it should be remembered that single-fluke anchors can be used instead for all mines, &c., moored on a span.

The necessary weights of sinkers for different circumstances have never received the attention which the subject deserved. Rules of thumb, founded on indifferent theory or none at all, were followed in the early stages of submarine mining, and were perpetuated because the time and attention of adepts have been occupied by what were apparently more important matters.

It is, however, a serious thing when mines walk about with their sinkers, and take up new positions which they are not intended to occupy; and a little theory (although "a dangerous thing") may, perhaps with advantage, be brought to bear upon this simple subject.

In order to arrive approximately at the necessary weights of sinkers for submarine mines, we must first fix upon a minimum coefficient of friction between a sinker and the bottom of the sea. The "R.E. Aide Mémoire," vol. i., sec. 73, states: "To move a stone along a rough chiselled floor requires $\frac{2}{3}$ of its weight." In other words, the coefficient of friction, under these circumstances, is $\frac{2}{3}$. But a sinker should be provided with projecting feet or claws which carry the whole weight on a few points, and as the weight is always sufficient to make the claws cut into such a surface as a chiselled stone floor, it is certain that even in the unfavourable and rare occurrence of the sea bottom being flat rock, the coefficient of friction would be something in excess of $\frac{2}{3}$. In most situations the bottom yields to the weight of the sinker, which thus becomes imbedded; and when this occurs the subsequent resistance to horizontal motion must be greatly increased. Assuming, however, that the minimum coefficient of friction between a sinker and the sea bottom is $\frac{2}{3}$, and this is certainly erring on the safe side of the truth, a few calculations will give the necessary weight of sinkers for contact and other buoyant mines.

Firstly, let us find the weight W of a sinker required to moor a spherical buoyant body possessing an available buoyancy B lb., a diameter D ft., and a diametric area A sq. ft., in a current of V knots

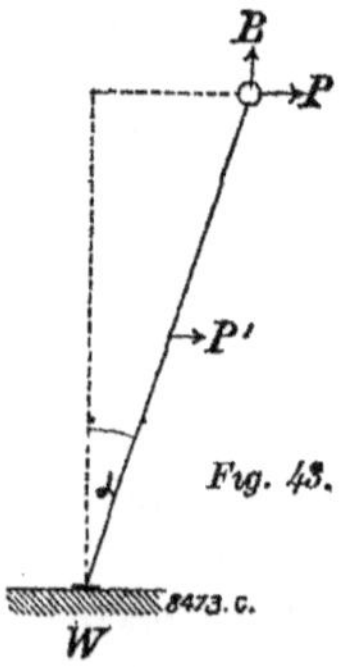

Fig. 43.

per hour, the size of the mine and the weights carried being such that the deflection of the system from the vertical is a.

Also let A^1 represent the diametric area of the cross-section of mooring line and electric cable (or of the electric cable alone, if the body be moored by the cable).

Then, if P be the side pressure produced by the current on the buoyant body, and if P^1 be the side pressure on the mooring line and cable, $P = 1.03\ V^2A$, and $P = 2.85\ V^2A^1$ (Froude).

For simplicity let P^1 be transferred so that its theoretical point of application is at the mine, the same result being obtained when $\frac{P^1}{2}$ is so applied.

Then
$$B = \left(P + \frac{P^1}{2}\right) \cot a$$
$$= V^2 \cot a\ (1.03\ A + 1.42\ A^1).$$

But
$$W - B = \text{the weight of sinker on the bottom,}$$
$$\therefore \tfrac{3}{8}\ (W - B) = \text{the horizontal force required to move it,}$$

and this should be $> P + \frac{P^1}{2}$ by an amount sufficient to withstand any additional strain which may be put upon the system.

During the mooring operations it is often necessary that a large boat shall hang on to the mine, and the resistance so offered to the current must be allowed for, and will, moreover, provide a margin of safety as regards the weight of the sinker. Assume that the resistance added by the boat so moored is equivalent to that of a totally submerged sphere with a diametric area of 10 square feet (it would be a large boat to offer such a resistance) the equation for weight of sinker then becomes

$$\tfrac{3}{8}\ (W - B) = V^2 \cot a\ (1.03\ A + 1.42\ A' + 10).$$

The same formula is applicable to a mine moored on a span with two sinkers, for if the weights be so adjusted that the down-stream line is just slack when the current is running its strongest, the result obtained from the equation will be correct; and if the buoyancy of the mine be greater than necessary, neither line will ever be slack, some of the buoyancy being supported by the down-stream sinker. If therefore the equation be applied to find the weight required for the up-stream sinker and the buoyancy held down by the other stream sinker be neglected, the result must evidently be on the safe side of the truth. If the tide run with equal velocity in the other direction, and if the angle a for each mooring line be approximately the same, the sinkers should be of equal weight.

Let us apply the formula to find W of each of the two sinkers used with a 42-in. spherical mine, containing a charge of 500 lb., and supporting 125 lb. of cable and mooring line in 21 fathoms, the whole arranged as previously described and illustrated; then the current velocity being 2 knots per hour,

$$B = 987^* - 625 = 362 \text{ lb.}$$
$$a = 45 \text{ deg.},\ A = 9.6,\ A^1 = 11,\ V = 2,$$

* See Table XX.

and

$$W = 362 + \tfrac{2}{3} \times 4\ (9.9 + 15.6 + 10).$$
= 575 lb. in salt water.
= 671 lb., say 6 cwt., in air.

Similarly for a $2\tfrac{1}{2}$-knot current,

$$W = 362 + \tfrac{2}{3} \times 6.25 \times 35.5,$$
= 696 lb. in salt water,
= 811 ,, say $7\tfrac{1}{4}$ cwt., in air,

and

W = 981 ,, ,, $8\tfrac{3}{4}$,, for a 3-knot tide,

and

= 1183,, ,, $10\tfrac{1}{2}$,, ,, $3\tfrac{1}{2}$,,

Applying the formula to the large 48-in. buoyant spherical mine in the same depth of water, but in currents ranging from 4 knots to $5\tfrac{1}{2}$ knots, we have: $A = 12.57$, $B = 1472 - 625 = 847$, and the rest as before. Then for a 4-knot current,

$$W = 847 + \tfrac{2}{3} \times 16\ (12.95 + 15.6 + 10).$$
= 1771 lb. in salt water.
= 2066 ,, say 18 cwt., for 4 knots.

Similarly,

W = 2354 lb., say 21 cwt., for $4\tfrac{1}{2}$ knots.
= 2673 ,, 24 ,, 5 ,,
= 3028 ,, 27 ,, $5\tfrac{1}{2}$,,

These heavy weights speak forcibly against the employment of buoyant mines of any description in swift currents, if such can by any possibility be avoided. Not only must the gear be heavy, but the continued strains and chafing are apt to damage the insulation of the electric cables and do other mischief.

Applying the Formula to Electro-Contact Mines, we find that the values for W come out thus:

When a spherical mine 3.25 ft. in diameter is loaded, primed, and moored in the service manner in water 11 fathoms deep, running 4 knots, its maximum efficiency is obtained when $a = 24$ deg. (see example on foot of page 61); also $B = 630$ lb., $A = 8.3$ square feet, $A^1 = 7$ square feet. Then by formula

$$W = 630 + \tfrac{2}{3} \times 4 \times 4 \times 2.25\ (1.03 \times 8.3 + 1.42 \times 7 + 10).$$
= 2169 lb. in salt water.
= 2530 lb. in air.
= $22\tfrac{1}{2}$ cwt.

Similarly, if the same mine be moored in the same water, but velocity of current = 3 knots,

$$W = 630 + \tfrac{2}{3} \times 3 \times 3 \times 2.25 \times 28.5.$$
= 1045 lb. in salt water.
= 1219 lb. in air.
= 11 cwt.

Again, if this mine be moored in 21 fathoms it is efficient in a current of a little over 2 knots, as shown previously. B is reduced by 50 lb., the weight of additional cable and wire rope; A^1 is increased by 7 square feet for same reason, and a is reduced to 18 deg. (whose cot = 3.08) for reasons already given on page 61.

$$\therefore W = 580 + \tfrac{3}{2} \times 2 \times 2 \times 3.08\ (28.5 + 7).$$
$$= 1290 \text{ lb. in salt water.}$$
$$= 1405 \text{ lb. in air.}$$
$$= 12\tfrac{1}{2} \text{ cwt.}$$

Turning to the examples given of a contact buoy* (and small suspended mine) on two sinkers we find that in 21 fathoms the dead-weight supported by the buoy = about 150 lb., and if V = 4 knots a 3-ft. sphere of steel $\frac{1}{4}$ in. thick possessing a buoyancy empty of about 620 lb. is required and the weight of sinker.

$$W = 470 + \tfrac{3}{2} \times 4 \times 4\ (1.03 \times 7.06 + 10.3 \times 0.5 + 1.42 \times 13.4 + 10).$$
$$= 1352 \text{ in salt water.}$$
$$= 1577 \text{ in air.}$$
$$= 14 \text{ cwt.}$$

Similarly in 21 fathoms and $3\frac{1}{2}$ knots,

$$W = 470 + \tfrac{3}{2} \times 3\tfrac{1}{2} \times 3\tfrac{1}{2} \times 36.8.$$
$$= 1146 \text{ in salt water.}$$
$$= 1337 \text{ in air.}$$
$$= 12 \text{ cwt.}$$

Now it was shown that a $2\frac{1}{2}$-ft. sphere would do for 3 knots with other conditions as above.

$$\therefore W = 290 + \tfrac{3}{2} \times 3 \times 3\ (1.03 + 4.9 + 1.03 \times 0.5 + 1.42 \times 13.4 + 10).$$
$$= 757 \text{ in salt water.}$$
$$= 883 \text{ in air.}$$
$$= 8 \text{ cwt.}$$

Similarly in a 2-knot tide,

$$W = 290 + \tfrac{3}{2} \times 2 \times 2 \times 34.6.$$
$$= 497 \text{ in salt water.}$$
$$= 580 \text{ in air.}$$
$$= 5\tfrac{1}{4} \text{ cwt.}$$

Again, in 11 fathoms and $3\frac{1}{2}$ knots and the same sphere,

$$W = 340 + \tfrac{3}{2} \times 3 \times 3\ (1.03 \times 4.9 + 1.03 \times 0.5 + 1.42 \times 6.7 + 10).$$
$$= 669 \text{ in salt water.}$$
$$= 780 \text{ in air.}$$
$$= 7 \text{ cwt.}$$

The weights of the sinkers shown in Table on next page, which would be required in the majority of harbours, are from 6 cwt. to $10\frac{1}{2}$ cwt. for the buoyant observation mines, and from 5 cwt. to 12 cwt. for the contact mines.

* The five following calculations should be slightly modified (see foot-note to page 59), but the results would not be materially affected thereby.

TABLE XXVIII.—SINKERS FOR SPHERICAL MINES, &c.
Found by Formula

$$W = B + \tfrac{3}{2} V^2 \cot a \, (1.03\,A + 1.42\,A^1 + 10).$$

Description of Case, &c.	Diameter.	Buoyancy when Loaded and Moored.	Depth of Water in Fathoms.	Velocity of Current, Knots per Hour.	Diametric Area of Case in Square Feet.	Longitudinal Area of Cable and Mooring Line.	Angle between Mooring Line and Vertical.	Weight of Sinker (in Air) required.	Weight of Charge Apparatus, Cable, and Mooring Line.
	$=144\,t$	B.		V.	A.	A^1.	a.	W	
	ft.	lb.					deg.	cwt.	lb.
Buoyant mine, large, on two sinkers... ...	4	847	21	5½	12.57	11	45	27	625
Ditto	4	847	21	5	12.57	11	45	24	625
Ditto	4	847	21	4½	12.57	11	45	21	625
Ditto	4	847	21	4	12.57	11	45	18	625
Buoyant mine, small, on two sinkers... ...	3½	362	21	3½	9.6	11	45	10½	625
Ditto	3½	362	21	3	9.6	11	45	8¾	625
Ditto	3½	362	21	2½	9.6	11	45	7¼	625
Ditto	3½	362	21	2	9.6	11	45	6	625
Contact mines on single sinker ...	3¼	630	11	4	8.3	7	24	22½	200
Ditto	3¼	630	11	3	8.3	7	24	11	200
Ditto	3¼	580	21	2	8.3	14	18	12½	250
Contact mines on two sinkers ...	3	470	21	4	7.06	13.4	45	14	150
Ditto	3	470	21	3½	7.06	13.4	45	12	150
Ditto	2½	290	21	3	4.9	13.4	45	8	150
Ditto	2½	290	21	2	4.9	13.4	45	5¼	150
Ditto	2½	290	11	3½	4.9	6.7	45	7	100

It now remains to settle upon the best shape of sinker. A convenient form was designed by the writer in 1875, was adopted in 1878 by the Government, and has remained the service pattern up to date. Prior to 1878 the mushroom sinkers did not house one into the other, and they consequently occupied valuable space in store, or encumbered the ground round the cranes, or sinker platforms had to be built whence they could be rolled on to the trucks. Moreover, when transported from the central store to out stations, they took up more space on board ship and were not so easily secured in the hold as at present. The sinker is circular, and all between 6 cwt. and ½ ton can be made of the same diameter, viz., 2 ft. 2 in.; the different weights being obtained by different heights. It has a flat top (see Figs. 44 and 45) and a strong central 3-in. eye made of ¾-in. wrought iron. Near the circumference there are three triangular indentations, about 3¼ in. deep at the outside and narrow part, and sloping upwards to nothing at the inner and wider portion.

Each indentation is provided with a wrought-iron bar across the top of the outer opening, and these bars not only strengthen the sides of the indentations, but act as additional eyes for attaching chains, &c., to the sinker, as required, and keep the feet in position when the sinkers are housed. The bottom of the sinker is slightly concave, and has three feet cast upon it which fit into the three indentations on the top of a similar sinker below it. The service pattern has four feet and four indentations instead of three. A few improvements now suggest themselves to the writer.

1. The wrought iron should be $\frac{7}{8}$ in. in place of $\frac{3}{4}$ in.

2. It would often be convenient to connect two sinkers rigidly to form a single sinker of greater weight, and this can readily be provided for by simply leaving three vertical holes (see Fig. 44) through which wrought-iron bolts could be passed and secured by suitable nuts.

Fig. 44.

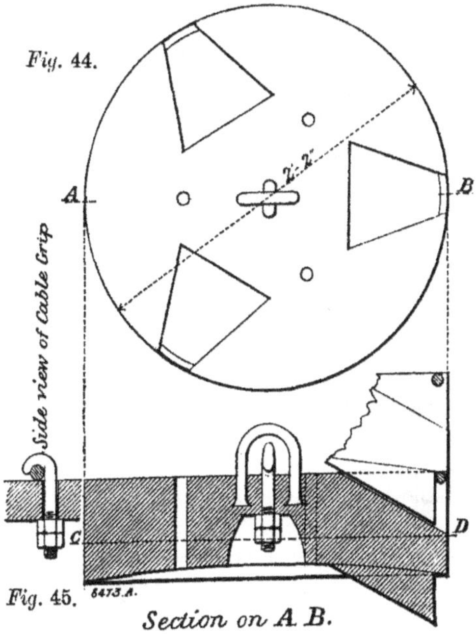

Fig. 45.

Section on A B.

3. As the system of mooring by the electric cable has been advocated in these articles, especially when two sinkers are employed on a span, provision should be made for securing the cable to a sinker in such a manner that it shall be firmly held at the centre of the sinker. When so used the central eye will not be required, and the cable can therefore be fixed by a hook passing through a hole from the central cavity in the bottom of the sinker, the nuts being put on by a box spanner.

Cast-iron sinkers of above pattern 26 in. in diameter weigh in air about 133 lb. per inch of depth, the average weight of cast iron being 444 lb. per cubic foot. They can be cast of any desired thickness between the limits of 5 in. and 13½ in., which give weights of 6 cwt. and 16 cwt. respectively. Beyond these limits sinkers of smaller or larger diameter should be employed, a good rule being that the depth shall range from about one-fifth to half the diameter. But if 26-in. sinkers be cast weighing 6 cwt., 8 cwt., and 10 cwt. in air, and the suggestion be adopted of bolting two together when required, we obtain a range from 6 cwt. to 20 cwt. at steps of 2 cwt. Thus:

$$6, 8, 10, (6+6)=12, (6+8)=14, (6+10), \text{ or } (8+8)=16,$$
$$(8+10)=18, \text{ and } (10+10)=20.$$

Such a range would probably meet all the usual requirements for the service.

If, however, it be considered advisable to provide for the range with still smaller steps, it could be done by adopting a further suggestion which is now made, viz., to manufacture the sinkers in three parts: top piece, that part shown above the dotted line C D in Fig. 45; bottom piece, that part shown below C D, and in the middle a number of iron discs of the commonest and cheapest ship-plate, say 1 in. thick, the discs being added or subtracted according to the weight of sinker required. When 20 cwt. has to be exceeded, a convenient diameter for cast-iron sinkers is 3 ft., and this gives 266 lb. per inch of depth, or double the weight of the 26-in. sinkers. A 20-cwt. sinker must then be 8.4 in. thick and a 30-cwt. sinker 12.6 in. thick. As, however, these sinkers are only required on rare occasions, they can be cast when required of the thickness to give the desired weight, and special patterns need not be stored.

It is seldom necessary or desirable to moor mines on sinkers weighing less than 5 cwt. in air. But light sinkers are required for marking buoys and other purposes, and it is convenient that the weight should then be adjustable between the limit of ½ cwt. and 2 cwt. or 3 cwt. This can be done by making the sinker (now suggested) of several iron discs and bracing them together by through bolts, the number of discs

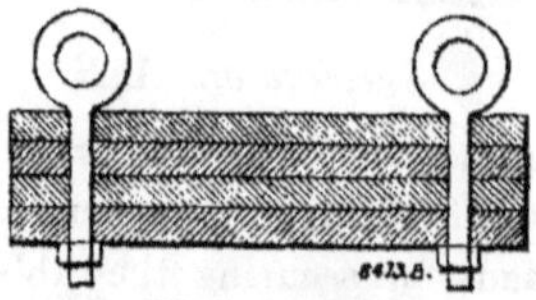

Fig. 46—47.

used giving the required weight. Discs 16 in. in diameter and 1 in. thick, made of wrought iron, weigh ½ cwt. each, and these dimensions are recommended. Two or more eye-bolts should be provided (as shown in Fig. 46—47) for mooring line, slip line, &c.

Mooring Lines.—These are made of flexible steel wire rope, the desiderata being strength united as far as possible with flexibility, lightness, small diameter so as to resist moving water as little as possible, and lasting power. Strength and flexibility are obtained by the employment of steel wires of small gauge, but durability in salt water is obtained best by wires of large dimensions. A compromise is therefore necessary. In the first place let us examine the working loads which may be brought upon such ropes. The outside limits of weight are practically given in the column marked W of the Table of sinkers, for these sinkers in many instances would have to be brought to the surface again by means of the wire ropes which connect the mines to them. Thus the large buoyant mine, Fig. 42 (page 84), say in 21 fathoms, the current running 5½ knots, must have 27-cwt. sinkers, and if they become imbedded in mud 50 per cent. additional tension might come on the wire rope before they would budge. Thus the tension might rise to as much as 50 cwt., say 2½ tons, and the breaking strain of such rope should therefore not be less than 15 tons.

Similarly, the smaller size of 500-lb. buoyant mines when moored in 21 fathoms, and a current velocity of 3½ knots, require two sinkers of 10½ cwt. each, or 16 cwt. as the safe load of the wire rope, say a breaking strength of nearly 5 tons. Also, the 3¼-ft. spherical electro-contact mines in 11 fathoms and a 4-knot current, if on a single mooring, require a sinker of 22½ cwt., and therefore a mooring line up to a safe load of nearly 34 cwt., say a breaking strain of 10 tons. Also contact mines on two sinkers in 21 fathoms, and 4-knot current, require sinkers of 14 cwt., and therefore a mooring line up to 21 cwt., or a breaking strength of nearly 6½ tons. But the large spherical mines on a single mooring are not recommended; and the large 500-lb. buoyant mines in deep and swift waters should only be rarely employed. For most situations, therefore, a breaking strength of 6½ tons is ample, and frequently one of 5½ tons is sufficient. A 2-in. steel wire rope may therefore be taken and accepted as strong enough for submarine buoyant mines in general, and 2¾ steel wire rope for the exceptional situations mentioned. The Table on the next page is for Mr. Bullivant's patterns.

The information given in the last column is important. If smaller drums be used, the wire rope is sure to be damaged, and the diameter of the drums on the steam winches of the mooring steamers, as well as those of the hand winches on the pinnaces, must be fashioned accordingly, and the gearing so designed that the winches will then work properly when the maximum strain is put on the wire rope. The larger the barrel the better, so far as the rope is concerned, but care must be taken to see that the machinery is then strong enough to turn it with the maximum load upon it. When the rope only passes over a sheave, the diameter of the sheave may be one-sixth less than the dimensions given. The 2-in. rope is very convenient for submarine mining. It consists of six strands round a hempen core, each strand containing twelve No. 19 B.W.G. galvanised steel wires. Its circumference and its weight are a little less than the figures given on the Table, and its strength is a little more. With a proof strain of about 4 tons it stretches about 1 in. per fathom. This rope will not deteriorate in store if kept well oiled, and Mr. Bullivant informs the writer that the 4½-in. patent flexible steel wire rope he supplied to the ship Lady Jocelyn in September, 1874, was, after it had been in use ten years, tested at Mr. Kirkaldy's public testing machine, and it actually took a greater breaking strain than was guaranteed at the date of supply.

TABLE XXIX.—PARTICULARS OF WIRE ROPES FOR MOORING LINES.

Circumference.	Weight per Fathom		Guaranteed Breaking Strain.	Diameter of Barrel or Sheave round which it may be Worked.	REMARKS.
	In Air.	In Salt Water.			
in.	lb.	lb.	tons.	in.	
3.0	7.0	6.1	18.0	18.0	The safe load should not exceed ⅙ the breaking strain.
2.75	5.5	4.8	15.0	16.5	
2.5	4.5	3.9	12.0	15.0	
2.25	3.75	3.3	9.0	13.5	
2.0	2.75	2.4	7.0	12.0	
1.75	2.0	1.75	5.5	10.5	
1.5	1.75	1.5	4.0	9.0	
1.25	1.0	0.9	2.5	7.5	Useful for mooring the marking buoys.
1.0	0.75	0.65	1.75	6.0	

It not unfrequently occurs that mines have to be raised and the length of their mooring lines altered as quickly as possible. The following arrangement recently designed by the author provides for such a contingency. The top of the mooring line, instead of terminating as usual in an eye to be shackled to the ring of the attachment chain or wire rope sling on the mine case, is merely cut to an end and served or crowned. The rope is then fastened to the ring by a double turn, and the end clamped to the standing part of the mooring line by a mechanical clip formed of two small iron plates with grooves across them for the wire rope, a central bolt and nut clamping the plates firmly together after the rope has been inserted.

Chain.—When a buoyant mine has to be raised it is not desirable to use the armoured electric cable for this purpose, although it ought to be strong enough if required. It is better to stopper a piece of chain along the cable of sufficient length to reach from the sinker to the surface, and thence over the sheave or joggle at the bow of the mooring steamer to the winch or capstan. The remarks just made concerning the necessary strengths of wire mooring ropes apply equally to these so-called tripping chains, and the following Table may be useful:

TABLE XXX.—PARTICULARS OF TRIPPING CHAINS.

Chain Cable, Short-Linked Size.	Weight per Fathom		Proof Strain.	Breaking Strain.	REMARKS.
	In Air	In Salt Water.			
in.	lb.	lb.	tons.	tons.	The chain should be galvanised. The width of each link should be $\frac{3}{5}$ of its length. The $\frac{3}{8}$-in., $\frac{1}{2}$-in., & $\frac{9}{16}$-in. chns. are those most usually employed.
$\frac{12}{16}$	30	25.0	10.1	15.1	
$\frac{11}{16}$	25	20.8	8.5	12.75	
$\frac{10}{16}$	21	17.5	7.0	9.5	
$\frac{9}{16}$	17	14.2	5.5	7.25	
$\frac{8}{16}$	14	11.7	4.5	6.0	
$\frac{7}{16}$	13.5	11.3	4.1	5.5	
$\frac{6}{16}$	10.35	8.9	3.75	5.0	

Shackles.—Before concluding these remarks on the mooring gear, a few words are necessary concerning a small detail that has given an infinity of trouble. An ordinary shackle with a split pin is unsatisfactory, as the split portion becomes rusty and frequently breaks. Perhaps, if made of steel, this defect may be rectified. A shackle with a screw pin is apt to unscrew and become unfastened by the constant swaying motion of a buoyant mine in a current, and if the pin be secured through its eye to the shackle by means of wire, the latter is not easily disengaged on a cold day by a man hanging over the bow of a mooring steamer with only one hand available. When picking up mines the necessity of a shackle that can be easily unfastened and yet that will not become unfastened unintentionally, became apparent.

A Scotch smith, M'Inlay by name, who was a sapper in the Royal Engineers, invented an arrangement which answered well. He secured and incorporated a small cross-pin with the shackle pin so that half of the cross-pin projected beyond the surface of the shoulder of the shackle pin, and he filed an indentation in the corresponding surface of the shackle, so that when the shackle pin was screwed up, the shackle itself was pushed back or bent until the cross-pin came opposite the indentation, and the spring of the shackle then caused it to fly back, making the whole secure. These shackles never come undone accidentally, and can yet be unfastened easily by a small marlinspike. Another excellent shackle is the invention and has been patented by Major R. M. Ruck, R.E. In this arrangement the pin is secured by means of an india-rubber washer, which engages the pin over a portion that is reduced in diameter. The washer keeps the pin in position when there is no strain on the shackle, but when the shackle is in tension a small catch at the end of the pin prevents it from slipping back. To open the shackle there must be no tension upon it, when the pin can be readily forced back by the thumb, and then pulled at the other end until the catch is brought up by a rubber washer. The shackle is closed by simply pushing the pin back. This shackle has recently undergone certain reputed improvements, but the foregoing description explains the idea underlying them, and if further information be required it can be obtained by writing to the makers, Messrs. Emerson, Walker, and Thompson Brothers, Winlaton, Blaydon-on-Tyne.

CHAPTER VIII.

Electric Cables for Submarine Mining Purposes.

General Remarks.—The manufacture of electric cables for submarine work has become a national industry, and one in which we have few competitors. Some of our foremost electricians are intimately connected with the great commercial companies thus evolved, and the matter is so thoroughly understood, and has so frequently been treated in various periodicals and publications, that it is unnecessary to do more than indicate the requirements of submarine mining, and any electric cable engineer will supply all the additional information that may be required.

Multiple Cable Cores.—In order to obviate the necessity of employing a large number of cables on a restricted area, it is desirable to employ multiple cables, and to lay them from the firing stations on shore to certain convenient points selected near to or on the mine-fields, to which points the single cables of the mines can be led and connected, each to one of the cores of the multiple. The disposition of the multiple cables can usually be so arranged that they need not cross an anchorage, and many of them can then be laid permanently, ready for war purposes. The most convenient multiple cables for employment in submarine mining are four-cored and seven-cored; they should be armoured for the sake of protection and strength; and there is no necessity for them to be brought round any small drum in the process of recovery, or what is termed "picking up."

Under such conditions gutta-percha is the best dielectric, and is therefore recommended for employment in the multiple cables for submarine mining, care being taken that it is stored in water from the time it is made to the time it is laid; and, consequently, that it is taken to the station in tanks, and never exposed to the direct rays of a summer sun in this country, or of a tropical sun at any season.

It will be seen, therefore, that multiple cables can be treated very similarly to the present deep-sea telegraph lines, and consequently that their general method of construction, and the manner in which they are laid, or recovered, may assimilate therewith.

Single Cable Cores.—It is impossible to deal so effectively with the single cables.

They must be exposed for a certain period during the process of connecting up the mines, cables, chains, wire ropes, sinkers, &c., on shore, and afterwards in the mooring operations. Also when a faulty mine is picked up, the single cable is again exposed during a repetition of these actions. Moreover, in picking up mines it is frequently necessary to take considerable strains on the single cables, and to hoist upon them by means of crabs or bollards revolved by power. Under such circumstances, the india-rubber covered core known as Hooper's is better than a gutta-percha covered core. This core can be stored dry on the drums as received from the makers, if a cool and dark place be available; but this description of storage for any great length of time is not so trustworthy as wet storage in cable tanks constructed for the purpose; or in the sea, in large coils, just below low-water mark.

Subterranean Cables on Shore.—*En passant* it may be remarked that Hooper's core unarmoured, but covered with felt tape, and by a layer of Manilla yarns, and preservative covering, with a plaited exterior of yarns, forms an excellent land-line cable for subterraneous work, to connect firing stations either for firing, or for telegraphic purposes. This form of cable can be made up very conveniently into four or seven-cored multiples, before being covered with the yarns and external plaiting.

Covered Wires for the Firing Stations.—Protected wires of a less costly nature will, however, answer every purpose for the connections in the firing stations, telegraph stations, and test rooms.

The Conductor.—The conductivity of electric cables for submarine mining depends to a great extent upon the sensitivity of the electric fuzes employed in the mines, and in succeeding chapters the employment of much more sensitive fuzes than those now used in the English submarine mining service will be recommended, it being possible to reduce the current required to fire a fuze from 0.9 of an ampère to 0.15 ampère, and to reduce the conductivity of the cable cores in like measure. Thus instead of the conductor offering only 5 or 7 ohms resistance per 1000 yards, it may have 40 ohms and be used efficiently in connection with a more sensitive fuze. It is, however, advantageous to possess a line of low conductivity to the mine when by any cause a leak has been developed in any portion of a core, and for this reason, if for no other, it is better not to reduce the conductivity too much. One-half of the reduction above mentioned can be taken with perfect safety, however, if fuzes of higher sensitivity be used. The conductivity resistance per 1000 yards may perhaps be fixed at about 18 or 20 ohms.

The conductor should be constructed of several wires twisted together, because greater elasticity in a longitudinal direction, as well as greater pliability, are thereby obtained. It is sometimes found that the conductor is broken by the strains thrown upon a cable during submarine mining operations, and the question arises whether a conductor made of twisted steel wires might not be employed advantageously instead of copper. For any given conductivity the steel would, however, be six times the cross-section, and 2½ times the circumference as compared with copper, and the amount of dielectric covering would be increased proportionally.

No experience having been obtained in this direction, it can scarcely be recommended except for experiment, but it appears worthy of trial as such, for the steel conductor being six times the weight, would be twelve times the strength of a copper conductor. The compound wire (an American invention) now manufactured in this country by Messrs. Siemens Brothers, would probably form an effective compromise.

The following figures are taken from the very complete Table published by Messrs. Walter Glover and Co., of Manchester, contractors to Her Majesty's Postmaster-General, &c. :

TABLE XXXI.—DIMENSIONS, &c., OF PURE COPPER WIRES.

B.W.G.	Diameter.	Yards per Pound.	Yards per Ohm.	REMARKS.
No.	in.			
10	.134	6.133	579.80	Stranded conductors weigh more and offer less electrical resistance than the amounts given by these figures.
11	.12	7.647	464.977	
12	.109	9.268	383.637	
13	.095	12.202	291.417	
14	.083	15.985	222.446	The electrical resistance increases with temperature by 0.21 per cent. deg. Fahr.
15	.072	21.242	167.302	
16	.065	26.063	136.425	
17	.058	32.734	108.624	
18	.049	45.863	77.528	The Table is for 60 deg. Fahr.
19	.042	62.425	56.960	
20	.035	89.892	49.797	The weights are calculated on the assumption that 1 cubic foot of pure copper weighs 555 lb.
21	.032	107.537	33.065	
22	.028	140.461	25.315	
23	.025	176.190	20.181	
24	.022	227.517	15.628	
25	.020	275.294	12.916	
26	.018	339.870	10.462	
27	.016	430.147	8.266	
28	.014	561.827	6.329	
29	.013	651.587	5.457	
30	.012	764.710	4.650	

It being advantageous to use a number of small wires, No. 30 B.W.G.

may be selected, and it will be found that twelve such wires will give the conductivity resistance limit already mentioned, viz., 18 ohms per 1000 yards. Thus $4.65 \times 12 = 55.8$ yards per ohm for such a strand, and 55.8×18, or 1004 yards, will therefore offer 18 ohms resistance at 60 deg. Fahr. These wires should be annealed and tinned before they are stranded.

In the gutta-percha covered cores for multiple cables, the conductor may be composed of a strand of seven No. 27 B.W.G. copper wires which possess the same conjoint resistance, 18 ohms, for a length of 1042 yards.

The Insulator.—Proceeding outwards, the dielectric employed need not be so thick, nor so carefully arranged to produce high resistance, as in submarine cables for long lines of telegraph, a comparatively very low insulation resistance being sufficient for our purposes. But it is important that the general arrangement of the covering dielectric shall be such as to insure permanency as far as possible. This cannot be shown by a high insulation test during or soon after manufacture, but the opinions of the best makers should be sought, and their advice followed in this matter.

The gutta-percha covered cores should be made with two coatings of gutta-percha prepared in accordance with Mr. Willoughby Smith's patent; and the india-rubber covered cores should be made in accordance with Mr. Hooper's patent. They are so well known that they need not be described here.

The main cost of a cable is due to its core or cores. I repeat that a cheap core of comparatively low insulation resistance will act efficiently for submarine mines if its permanency be carefully provided for, both chemically and mechanically, by a tough insulator and strong pliable covering. The insulation resistance is not a matter to haggle about; the permanency is. The weight of the insulation need not exceed 1 cwt. per knot.

Core Covering.—Proceeding outwards, the core of a single cable should be covered with a serving of india-rubber coated cotton tape, wound on spirally with a fair overlap. This should be covered by a braiding of several three-ply fine hemp twines, and the whole steeped in a protective composition.

The Armouring.—As pliability is necessary, combined with strength and durability, strands of galvanised steel wire should be employed. A strand composed of seven No. 19 B.W.G. can be recommended, with say one twist in about 2½ in. Twelve of these strands will cover the cable when laid on with one turn in about 8½ in. to 9 in.

The Outer Covering.—In order to prevent the steel strands from gaping and thereby exposing the interior core to the attacks of marine life, &c., an outer covering of braided hemp cords should be added. This braiding should not bind the armouring too tightly, or the conductor will be unduly strained and perhaps break when the cable is bent and in tension.

The cable should now be steeped in a preservative compound.

Dimensions, &c.—The cable will then be about $\frac{3}{4}$ in. in diameter; it will weigh about a ton per knot (in air); and its breaking strength will slightly exceed 5 tons. Portions of it should be tested for conductivity when subjected to its working strain of, say, 1 ton, and a clause should figure in the specification to that effect. Also the conductivity and insulation resistances of the whole length should be found during manufacture, at delivery, and periodically afterwards, records of same being kept. The greatest care should be taken to see that the steel wires are thoroughly galvanised, and that the galvanising is not cracked during the process of stranding. As far as the armouring is concerned, dry storage is of course preferable to storage under water. The supply and delivery can be had in 1-knot lengths.

Multiple Cables.—Returning to the multiple cables, the four or seven cores already described should be stranded, and then wormed and served with tarred jute yarn, round which are wound about fourteen No. 12 B.W.G. galvanised B. B. iron wires for the four-cored, and about sixteen ditto for the seven-cored cable, to form an armouring. A braiding of hemp cords, as in the single cable, is then added, and the whole steeped in a preservative composition. The four-cored cable weighs about 2 tons in air and 1 ton in the water—and the seven-cored cable does not differ greatly from it in these respects.

Shore Ends.—When cables have to cross rocks or shingle exposed to a heavy wash from the sea, it is desirable to employ an additional armouring for their protection. The cable so employed may be precisely similar to the multiple-armoured cables just described, with an additional serving of tarred jute yarn and a sheathing of No. 1 B.W.G. (= 0.3 in. in diameter) B.B. iron galvanised wires. About eleven such wires are required to cover the four-cored, and twelve to cover the seven-cored cables. This cable being stiff, heavy, and difficult to coil or manipulate, should be laid as soon as possible *in situ*. The manner in which cables should be tested periodically will be described hereafter in the paper on testing stores.

Crowning Cables.—Cables must be connected to the mines and to each other both electrically and mechanically. For this purpose the cable end should be made into a crown, a padding of spun yarn being wound round the armouring about 1 ft. from the end of the cable, and the wires turned back and whipped with binding wire and spun yarn. The miners should be taught to make these crowns in accordance with patterns of the proper sizes and dimensions for multiple and for single cables. They will then fit into the cable grips made in the mine cases, junction boxes, &c. The projecting foot of the core should be protected with a whipping.

Electrical Joints.—The electrical joints used in submarine mining are very similar to those used for underground telegraphs, and require no special notice. Whenever time is available they should be soldered; Fletcher's soldering apparatus being used on the mine fields. Cable ends when crowned can be connected by laying them together, with about 6 in. of overlap, and lashing with binding wire and spun yarn. The cores can then be connected electrically, and the whole covered with a strong bandage of canvas. But cable ends are usually connected by what are termed :—

Fig. 49.

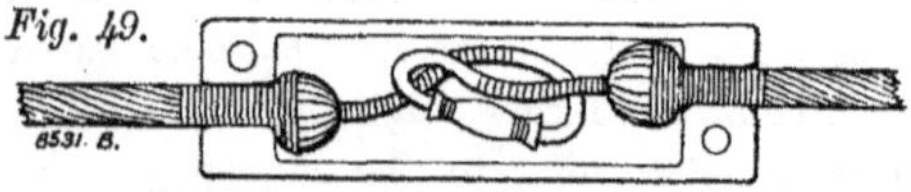

Box for connecting 2 Single cables Elec.

Fig 52.

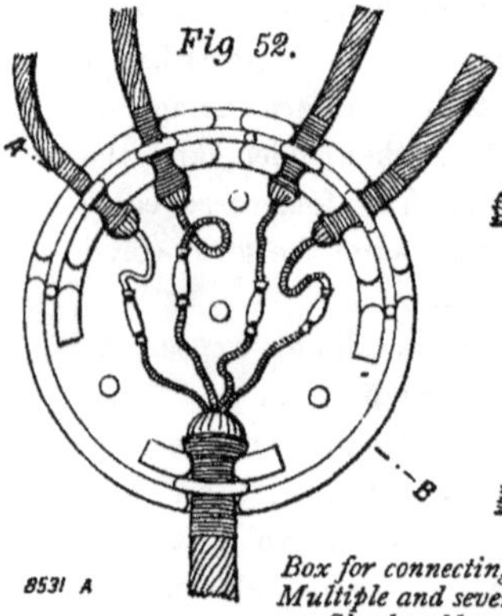

Box for connecting 1 Multiple and several Single cables.

Fig. 50. Box for connecting 3 Single Cables Elec.^d & a disconnector.

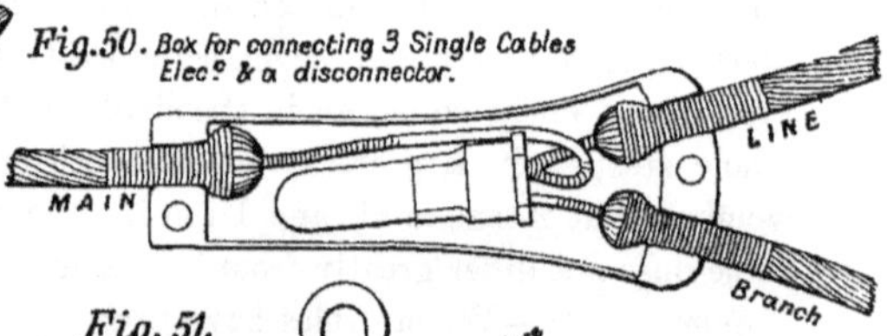

Fig. 51.
Section on A. B.

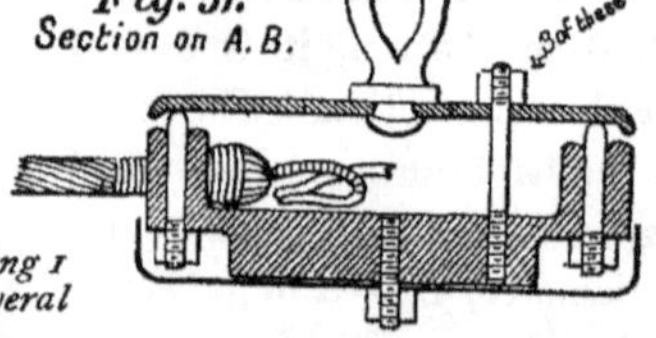

Fig. 48.

Grip hook

Connecting Boxes.—These can be made of cast iron, each in two half-pieces connected together by bolts and nuts, as shown on the sketch, Figs. 49 and 50. The bolts should be short; the heads small, and embedded in recesses in the castings; the nuts small, but deep, and also embedded as far as possible, so that the projections may not foul anything when the cable is paid off a drum or coil during the mining operations. The boxes for connecting multiple cables, four-core and seven-core, and for connecting a shore end with an ordinary multiple cable, may be similar to the above, but of suitable dimensions to grip these larger cables, and to hold the seven-core joints. It is not necessary to illustrate or describe them further.

It is sometimes necessary to connect three cable ends. The cast-iron box can then be made as shown on sketch (not the English pattern), see Fig. 50. It should be large enough to hold a small apparatus called a disconnector, to be explained hereafter.

Junction Boxes.—As explained, multiple cables lead to a number of single cables, one of the latter to each core of the former. These connections are made in what is termed a multiple junction box. It is best to have only one pattern, viz., for one multiple and seven single cables. The same box will then do very well for the multiple four-core cable and four single cables. It is often necessary to buoy the junction box both during and after the mining operations, and for this reason it should be somewhat heavy. The buoy can be comparatively small and be moored to a light line, which will bring up a chain from the bottom wherewith to weigh the box and its eight cable ends. A $\frac{7}{16}$-in. chain is not too strong for the work, and a $\frac{1}{2}$-in. chain in deep water. The multiple cable should be moored to a heavy sinker at a distance from the box of about twice the depth of the water. This prevents the system being dislodged when the box is raised for any purpose by a junction box boat in a tideway. If the human hand be opened, the thumb kept as far from the four fingers as possible, and the whole pressed down upon the table, keeping the arm vertical, it represents the system, the arm being the recovering chain, the thumb the main cable to shore, the four fingers four branch cables to the mines, and the palm of the hand the junction box. It is better to make the box heavy because the cables are then less liable to foul one another, or bottom obstructions. It makes it a little more difficult to raise the system, but a foul cable is far worse. The box should probably not be less than 2 cwt. when empty and on shore. Its shape should be circular in plan, as this gives the largest interior space for a given periphery, and also because it offers no corners to knock holes in the boats when raised in a seaway. An arrangement of the kind is illustrated in Figs. 51, 52. The plan shows the box with the wrought-iron cover removed. The cables are secured by grip hooks (Fig. 48), the nuts for same being readily got at. The sheet-iron disc covering the bottom is not absolutely necessary, but it protects the bottom of the grip hooks from blows, and gives a neater appearance. The thick wrought-iron lid is secured to the casting by three strong studs and nuts. The whole is recovered by the ring in the centre of the lid. This is not the English pattern, but is a decidedly better and stronger arrangement. When the branch cables lead to electro-contact mines, the main cable is a single cable, and the junction box is made larger in order to hold certain apparatus by which each mine is cut off from the system when fired, and by which tests can be taken when the box is raised. A new and improved apparatus of this kind recently designed by the writer will be described hereafter.

Cable Entry to a Mine.—When a cable is taken into a mine it is connected to one leg of the entrance plug electrically, and the cable crown is gripped by a cast-iron dome screwed down upon it at the same position that one of the grip hooks occupies on the sketch.

CHAPTER IX.

ON ELECTRIC FUZES.

THE efficiency of a submarine mine fired and controlled by electricity depends to a great extent upon the constancy and reliability of the electric fuze that is employed. Electric fuzes can conveniently be divided into two classes, viz. :

1. High-resistance fuzes.
2. Low-resistance fuzes.

The former can be ignited by small currents of high potential; the latter by larger currents, the potential required being much lower because the electrical resistances in the fuzes are insignificant.

One of the best papers ever written on this subject came from the pen of Captain (now Major-General) E. W. Ward, Royal Engineers, *vide* Paper XVI., vol. iv., of the R. E. Professional Papers, published 1855, where we find that the action of certain high-resistance fuzes then in use depended upon "the combustion of a compound, which seemingly is a sulphuret of carbon and copper," and that these fuzes were "the invention of Mr. Brunton, of the Gutta Percha Works in the City-road. This company had been in the habit of what is familiarly called vulcanising the gutta-percha which covered the wire, to render it pliable even in the coldest temperature, and this led to the discovery of the fuze in question. By the vulcanising process sulphur, and, I believe, carbon, became incorporated with the gutta-percha. . . . . These two act on the inclosed copper wire, and in process of time produce on its surface a species of sulphide, portions of which, when the wire is withdrawn, remain adhering to the inner surface of the gutta-percha covering. This inner surface . . . . has now a feeble power of conduction given to it by means of the minute particles of sulphide of copper and carbon. The conducting power is, however, very feeble, and, seemingly, in no two portions the same; but whatever the amount of resistance may be, if it can be overcome sufficiently to circulate such a force as will ignite the sulphur and carbon, the desired effect is obtained."

In Lieutenant-Colonel (now Major-General) Stotherd's "Notes on

Defence by Submarine Mines," published 1873, the Beardslee fuze with graphite priming and the Austrian and the Prussian fuzes with priming of ground glass and sulphur are described, and then the English service submarine mining fuze of that date as follows: "Another similar form of fuze is that invented by Mr. Abel, F.R.S., chemist to the War Department. This fuze was devised and experimented with extensively in 1858, and the above more recently designed fuzes, viz., Beardslee's, the Austrian, and the Prussian, are based upon the principles first applied to that fuze.* It has been modified since its first invention in a few details. . . . The priming of the original fuzes consisted of 10 parts of subphosphide of copper prepared by a special method, 45 parts of subsulphide of copper, and 15 parts of chlorate of potassa. These proportions of the ingredients are, however, now varied so as to furnish fuzes of different degrees of conductivity and sensitiveness to suit different purposes."

The fuze is then described in detail. It remained the service fuze for submarine mining for several years, and was finally abandoned for several reasons, one being the admitted danger that mines primed by any high-resistance fuzes might be accidentally exploded during magnetic or electric storms, either by induced currents or by the electric discharge of the cables at such times. It appears, also, from a sentence in General Abbot's book, that a fear of the effect of induced currents, however produced, "had much to do with their" (similar fuzes) "ultimate exclusion from" the American submarine mining service.

Low-Resistance Fuzes.—The ignition of gunpowder by a thin platinum wire heated by the passage of an electric current had already "been long in use" when Captain E. W. Ward, R.E., investigated the subject in 1855; but to him we are indebted not only for some simple and efficient instruments which have been in use ever since, but also for treating the matter scientifically, in precise mathematical terms. Moreover, the form of wire fuze adopted by him was retained in our service for a number of years, and the men taught to make them.

A small block of soft wood has a small rectangular cavity cut out of one side, and two No. 16 percha-covered wires are bared at the ends and passed through small cross-holes made in the block. The platinum wire is then soldered across them, the cavity filled with meal powder, and a thin wooden lid screwed down upon it. See Figs. 53 and 54. It is described, both on account of the history of the low-resistance fuzes, and because it is still useful as a method of improvising fuzes for

* This is not correct. The principles were first applied in the Brunton fuze, described.

gunpowder mines. In 1868 the following improvements were made by the author, who at that time was acting as assistant instructor of telegraphy at Chatham.

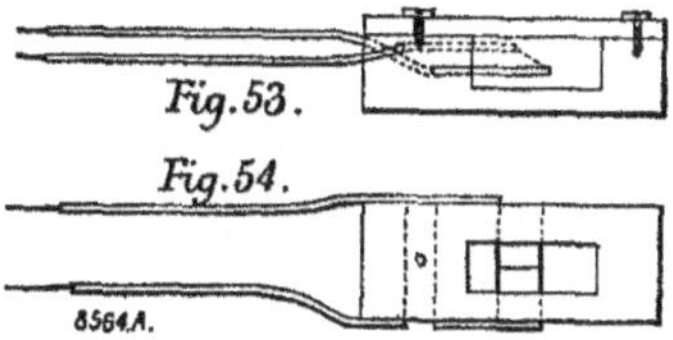

Fig. 53.

Fig. 54.

8564.A.

The two No. 16 wires were replaced by two thick wires passed through a block of wood shaped like an ordinary medicine cork, the platinum wire was soldered across their ends, and the sensitivity greatly increased by a wisp of fibrous gun-cotton wound round the platinum wire. Also the fuze was converted into a detonator by the employment of a charge of mercurial fulminate contained in a small tin cylinder, as shown in Fig. 55.

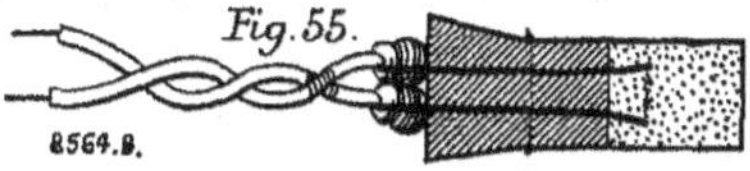

Fig. 55.

8564.B.

The employment of fulminating mercury to convert a fuze into a detonator had already been adopted in connection with the high-resistance submarine mining fuze already described, but up to the time when the other above improvements were made in the low-resistance fuzes, it was generally held by experts in England that these fuzes were inapplicable to submarine mining on account of their want of sensitivity, and the large battery power necessary to fire the mines at any distance from the battery. Experiments with the above fuze soon proved that this view was erroneous, and although the inventor was ordered abroad in the middle of them, they were continued, and the detonator was very favourably considered. In 1871, when the first edition of Captain Stotherd's "Notes on Submarine Mining" was published by the Royal Engineer establishment, the fuze was described by the name of the present writer, although strictly it was the Ward fuze improved. The next important improvement was made by Captain Fisher, R.N., when in command of the Vernon Torpedo School. He carried out a number of experiments with different alloys in the bridge, and finally selected platinum-silver, which is still employed in the Royal Navy in preference to other alloys. Captain Fisher's investigations may be said to have nearly perfected the fuze as now employed in the Royal Navy, but an improved wire was not adopted by the Royal Engineers until three years later, by which time experiments had been made with different wires at the Chemical Laboratory at Woolwich, as recorded on an important memorandum to the Society of Telegraph Engineers by Professor Abel, May 13, 1874. The outcome was similar to Captain Fisher's experiments, and confirmed those of General Abbot, which had then come to a termination, but had not been published, viz., that German silver is liable to corrosion ; that platinum-silver is superior in this respect, but difficult to draw into a fine and uniform wire ; that platinum-iridium fulfils the required conditions best, being easily fused, easily drawn, and safe against corrosion, also larger in diameter than platinum wire of same resistance, and therefore offering more surface to the priming. General Abbot's excellent report on fuzes treats the subject in a very thorough and scientific manner. He recommends that :

1. The insulated conducting wires for all fuzes should be formed of tough and flexible copper about 20 B. W. G., of equal length, say 5 in. and 7 in., and covered with a closely woven wrapping of cotton thread coated with paraffin, or with beeswax, resin, and tar boiled together. The employment of gutta-percha or india-rubber is not recommended, owing to deterioration after lengthened storage.

2. The plug of the fuze should be hard, strong, and a good non-conductor of electricity. Beechwood, kiln-dried, and coated thickly with Japan wax, is recommended in preference to other materials. The following form has been adopted into the American service. It is in three parts. First, a cylinder, 0.25 in. in diameter and 0.7 in. long, grooved longitudinally on opposite sides to receive the wires. Entirely round the middle is a cut 0.05 in. deep and 0.15 in. wide. The wires are carried up the horizontal grooves for half the length of the cylinder, then half round by the canelure, and up the remainder of the cylinder on opposite sides. The inside ends, about 0.1 in. long, are then bared and scraped. Second, a hollow cylindrical cap closely fitting above with a stout shoulder at one end, against which the solid plug abuts when it is forced into the cap, thus leaving a smaller hole for the passage of the free ends of the insulated wires. This leaves a small chamber round the bridge for the priming, about 4 grs. of mercurial fulminate, and the chamber is closed by a paper disc.

3. The detonating cap of the American fuze is made of copper punched into cylindrical form to fit the cap closely. It contains 20 grs. of mercurial fulminate, the total priming therefore being 24 grs.

4. The American fuze is 1.4 in. long and 0.4 in. in diameter. It is waterproofed with a coating of Japan wax.

There are several patterns of English fuze or detonator, the term fuze being applied by us to a fuze with a gunpowder bursting charge, and the term detonator to a fuze with a bursting charge formed of some detonating composition, and mercurial fulminate is found most suitable for this purpose. As General Abbot experimented with more than one form of English detonating fuze, and his report has been published, they cannot be considered as secret or confidential. Moreover, their important points have been made the subject of scientific papers read in public by Sir Frederick Abel, who is chiefly responsible for the specifications and patterns governing their manufacture at the Royal Arsenal.

Our detonating fuzes are now made with an ebonite head, in which two strong copper wires or rods are firmly imbedded. To them are soldered the leads, which are composed of multiple copper wires gutta-percha covered. An inner sheet-metal cylinder covers the pillar ends and bridge, a hole being left in the cylinder at the bottom by which it is filled with a mixture of finely powdered gun-cotton and mealed gun-powder in equal parts. An outer cylinder of thin metal prolonged into a small quill-shaped chamber contains the bursting charge of 25 grs. mercurial fulminate. The entire arrangement is shown on the sketch, Fig. 56.

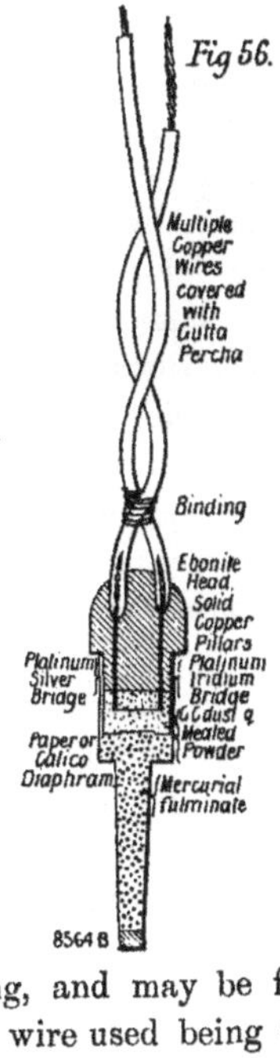

Fig 56.

The bridge is ¼ in. long, and may be formed of platinum silver, 33 per cent. platinum, the wire used being 0.0014 in. in diameter, and weighing 2.1 grs. per 10 yards. Its resistance (cold) is then 1.6 ohms, and the firing current about 0.27 ampère. When a platinum-iridium (about 10 per cent. iridium) bridge is employed, the wire being 0.0014 in. in diameter, or weighing 3.4 grs. per 10 yards, it has an electrical resistance (cold) of 1.05 ohms, and is fired by a current of about 0.17 ampère.

When firing a number of charges in divided or "forked" circuit, it is considered by experts in the Royal Navy that the fuze bridge should be composed of an alloy which melts at a low temperature. Platinum silver has a lower fusing point than either platinum or platinum-iridium. It has therefore been chosen for use in the naval fuzes and detonators used for firing broadsides and lines of countermines, for which purposes divided circuit has been adopted, no doubt for some good and valid reasons. For submarine mining purposes the charges can always be fired in series, and the best fuze then seems to be one in which the bridge is composed of platinum-iridium. The submarine mining fuze employed in our service has a much lower resistance than that given above, the wire being platinum-iridium (10 per cent. iridium) 0.003 in. in diameter, and weighing 1.55 grs. per yard. The firing current is about 0.9 ampère, and its fusing current about 1.65 ampère. Its resistance (cold) is 0.325 ohm, and 0.74 ohm at fusing point. The employment of a fuze with this low sensitivity is necessary for the particular arrangements employed in our service for firing, testing, and controlling the mines, but as these arrangements are secret and confidential, and as moreover they are not recommended by the present writer for general adoption for war purposes on account of their intricacy, and the difficulties consequently encountered in training men to satisfactorily perform the various operations, to say nothing of the utter impossibility to fill their places promptly in the event of numerous casualties occurring, the adoption of a more sensitive fuze or detonator, and of a much simpler general arrangement in the test and firing stations, is and will be recommended on these pages. As regards the temperature of the wire necessary for ignition, and the best priming to employ around the wire bridge, General Abbot makes some very pertinent remarks. He notes that gun-cotton flashes at 428 deg. Fahr. and mercurial fulminate at 392 deg. Fahr., but that the latter being a better conductor of heat lowers the temperature of the bridge more rapidly, and requires a slightly stronger current to fire the fuze. Nevertheless he prefers the fulminate priming on account of its greater uniformity in results, due probably, he says, to its greater weight bringing it more thoroughly in contact with the bridge. He adds, "While not a single instance of failure has been recorded among the thousands of fulminate of mercury" primed "fuzes used in these investigations, several gun-cotton" primed "fuzes have failed by the deflagration of the wire without the ignition of the gun-cotton priming." It may, however, be accepted that the gun-cotton priming is very efficient and quite reliable when care is taken to insure good packing. Some authorities recommend the wisp of gun-cotton (already referred to) improved by soaking it in collodion; but it is difficult to remove all traces of acid from long staple gun-cotton, and the other devices described are consequently preferable. The heat theoretically produced in the wire bridge of a fuze may be considered as follows:

If H be the number of units of heat developed by
C an electric current in ampères through
R a resistance in ohms in
T seconds of time,

$$H = C^2 R T.$$

If ρ be the specific resistance of alloy used,
l be the length of bridge, and
r its radius,

$$R = \rho l \div \pi r^2.$$

$$\therefore H = C^2 T \rho l \div \pi r^2.$$

But the rise of temperature x of the fuze wire varies directly as H and inversely as S the specific heat of the alloy, and the mass of the wire $l \times \pi r^2$.

Consequently,

$$x = H \div S l \pi r^2.$$
$$= C^2 T \rho l \div S l (\pi r^2)^2.$$
$$= C^2 T \rho \div S \pi^2 r^4.$$

The whole mathematical theory of the bridge of a wire fuze is examined with great care by General Abbot in his report, page 227 *et seq.*; but the above short method is sufficient to indicate the chief points of theoretic importance, viz.:

1. That the alloy should possess the highest possible specific resistance;
2. Combined with a very low specific heat;
3. That the cross-section of the wire should be as small as possible consistent with strength.

But another point does not enter into the formula, viz., loss of heat by conduction through the priming, and by conduction through the metallic pillars of the fuze. The former has already been alluded to. The latter can be met by making the bridge of sufficient length, a wire of large section requiring a longer bridge, which, however, should never be longer than is necessary for producing the required sensitivity.

In the American service the maximum current required for working their automatic arrangements on shore for firing the mines is 0.15 ampère, and as the bridge heat varies as the square of the current, allowing 10 as a margin of safety, the firing current has been fixed by them at $\sqrt{10\,(0.15)^2} = 0.47$ ampère. For fixing the bridge wires they use a solder which melts (with resin flux) only at a high temperature. The pillars are notched, put on the plug, tinned and gauged to the exact length of bridge. The wire is then soldered on, and the pillars bent slightly inwards, so as to take off any tension on the bridge.

When automatic arrangements are not employed at the firing and testing stations, more sensitive fuzes than those adopted for the American submarine service can be employed.

Disconnecting Fuzes.—Various electrical devices have been proposed and adopted by different nations for automatically cutting off a branch cable when a mine at the end of it is exploded. When low-resistance detonators are used in the mine an excellent arrangement is to use a similar low-resistance fuze as the cut-off. No detonating charge is employed, but a minute charge of pressed meal gunpowder is placed in a small tube having its end just below and pointing on the bridge. The rush of gas caused by the ignition of this small charge breaks the bridge wire, if the current has not already done so.

The simultaneous ignition of two or more fuzes on continuous circuit can be assured when the fuzes are well designed and made chemically and mechanically similar to one another, and the current is sufficient. Their minimum firing current is then very uniform. But a firing current should be used in practice that is at least equal to their fusing current, in order that all the fuzes may be ignited simultaneously with certainty. The explosion or detonation of the charges surrounding them does not then affect the problem, because "the time needed to raise the temperature of the bridge to the requisite degree" is thereby "made less than the minimum required to perform the mechanical work of explosion" (Abbot). When a weaker current is used, the time required to heat a slightly insensitive fuze may be less than the time occupied by the explosion of a neighbouring charge, and a blind charge be thereby produced. These remarks apply equally to the more complex problems connected with rock blasting, in which large numbers of fuzes are ignited simultaneously.

In submarine mining we have two fuzes in a mine and one in the disconnector, three in all. Also several mines are sometimes fired simultaneously, perhaps as many as five mines, when there would be ten fuze detonators in continuous circuit.

Extremely Sensitive Wire Fuzes.—For some purposes connected with submarine mining extremely sensitive wire fuzes may be employed with advantage. Such fuzes and detonators are employed in the Danish service for certain purposes, but their manufacture is a secret. General Abbot has, however, investigated the matter with his customary care and accuracy. He employs very fine wires composed of an alloy of platinum and some other metal, such as silver, that can be removed by oxidation, &c.

Short lengths, 0.44 in. long, of the wire are soldered to copper wire terminals, and are bent outwards into loops. They are then covered with wax, except about 0.08 in. at the centre, which is subjected to the action of nitric acid, and the silver removed chemically. The final result is a short length of platinum wire as small as 0.0002 in. in diameter, which when used in a carefully primed fuze with the above bridge length can be fired by a current of about 0.04 ampère. Adopting Abbot's co-efficient of safety, viz., $\sqrt{10\,c^2}$, where c is the safe current that can be passed through a fuze which can be fired by a current $= \sqrt{10\,c^2}$, and equating this with the minimum current which will fire the platinum-iridium fuze described at top of page 114; we have $0.17 = \sqrt{10\,c^2}$. Hence, $c = 0.17 \div \sqrt{10} = 0.054$ ampère. But 0.04 ampère is sufficient to fire the very sensitive fuze. Consequently the latter can be fired by a current which is quite safe to send through the former.

The following curiosity among fuze designs is termed the "Browne compound fuze, high and low tension," and consists of a platinum wire fuze at one end and a high tension fuze at the other, with the poles connected by two 5-ft. lengths of No. 22 copper (insulated) wire wound round the fuze. The firing leads are connected to the pillars of the high tension portion of the arrangement. A current of 0.85 ampère through the main leads fires the wire fuze, and a current of frictional electricity fires the high tension composition; thus proving what nonsense is accepted concerning the absolute protection afforded by lightning conductors. Mr. Browne's composition consists of mercurial fulminate four parts, sulphuret of antimony one part, and powdered antimony three parts.

In conclusion, fuzes for submarine mining, whatever may be the pattern selected, should be so designed as to be mechanically strong, chemically permanent, electrically uniform. Moreover they should be manufactured with the greatest care, stored in a suitable manner, so as to protect them from damp, and tested periodically to prove that they remain in an efficient condition.

The wire employed in the manufacture of fuzes is generally supplied from a well-known contractor, and the specification governing such supply should state the weight of a given length taken hap-hazard in grains; its resistance at 60 deg. Fahr. in ohms; the fusing current in ampères when passed through, say $\frac{1}{4}$ in. length; the firing current in ampères when passed through same, primed in the required manner; and the chemical composition of the wire.

The periodical tests of the fuzes should consist of resistance tests and sensitivity tests. Limits should be laid down to govern these tests. The boxes should be marked, and records kept of the tests and dates.

CHAPTER X.

Electrical Arrangements on the Mine Fields: In the Mines, Circuit-Closers, &c.

Observation Mines.—The electrical arrangements in connection with observation mines may be of the simplest possible form, viz., insulated conductor from firing station, through fuzes in mine (or mines, two or more being sometimes fired simultaneously), to "earth," in the sea, and thence by "earth return."

The electrical resistance is then the only test which can readily be taken to judge of the efficiency or otherwise of the system, and it is probably sufficient; but opinions differ on this point, many experts considering that an apparatus should be placed in the mine (or the end mine, if more than one), which will indicate the efficiency of the system more thoroughly.

A small electro-magnet is probably the best apparatus to employ, the movement of its armature by a small electric current, that can be sent through the fuzes safely, giving indications at the firing station that the system is in working order. These indications can be seen by means of a galvanometer, or heard through a telephone. The apparatus should be so arranged that it will cease to act properly when wet, and it should be placed at the bottom of the small chamber containing the priming charge of the mine. Should this chamber leak, it is then at once discovered.

Any electrical engineer can design such an instrument in half an hour, so no more need be said.

Mines with Circuit-Closers.—But an instrument may be designed which is not only capable of testing the mine or circuit-closer, but also of controlling the electrical connections therein. The author believes that he was the first to propose such an arrangement, on the 9th March, 1871. The following were the words used in the memorandum: "A quantity battery is in connection with the upper plate of a switch or commutator. A tension battery connects with the lower plate; and the third plate on which the axis of the switch handle is fixed connects with the cable. The mine may be incorporated with the circuit-breaker or be below it, and separate, but in either case the tension fuze is kept insulated from the cable. One pole of the fuze is put to earth, and the other is in connection with the metallic uprights of Mathieson's inertia circuit-closer, modified so as to be also a circuit-breaker. Inside this is an arrangement consisting of a coil of rather thick wire wound round two soft iron cores with an armature pivotted centrally between them. On the axis of this armature, and fixed to it, is an ebonite disc, across which a metallic wire is led. A fixed metallic point in connection with the shore cable presses against a small angular return on the circumference, and near the bottom of the disc. At or near the other end of the disc wire are two metallic points, one in connection with the standard of the inertia bob, and the other with the fuze, or with the metallic uprights already referred to, these being insulated but in connection with one pole of the fuze. The armature is kept open by a small spring, or by a preponderance in the disc. The action is as follows: If required to fire by contact, the tension battery is switched to cable and a constant current passes along line to the circuit-breaker. This current, however, is not of sufficient quantity to form an electro-magnet and attract the armature. As soon as a ship strikes the circuit-breaker the bob leaves dead 'earth,' and strikes against the uprights, and the tension current is thus switched through fuze and fires it. Again, if it be required to fire the mine by judgment, two motions of the switch handle in quick succession are necessary."

The quantity battery thus attracts the armature, and the tension battery fires the mine before the armature returns to its normal position. This arrangement gave power to test line for insulation, and also for movement of armature; but this was not noted on the memorandum.

There were several defects in this arrangement, and it is only described as a matter of historical interest.

On the following year Captain (now Lieutenant-Colonel) R. Y. Armstrong, R.E., invented a much better arrangement, which was eventually adopted into the English service.

Its present more elaborate and perfected form is a secret, but the broad principles are described in Stotherd's "Notes on Submarine Mining," published in 1873.

It consists of a polarised electro-magnet, its armature being pivotted centrally between the four poles of the electro-magnet, which latter is wound by a coil of thick wire offering small resistance, and by a separate coil of fine wire offering a high resistance.

One end of the thick coil and one end of the fine coil are generally connected to "line," whether the apparatus be placed in a mine or in a detached circuit-closer. The other end of the thick coil is generally connected to a stop against which the armature impinges and rests when attracted to the electro-magnet. The other end of the fine coil is generally connected to earth. The armature is generally connected to earth, and in the mine this path generally traverses the fuzes. A small positive or negative current from the testing station gives a small deflection on a low resistance galvanometer, and an increased positive current gives a large deflection due to the armature in the mine being attracted to the stop; also an increased negative current produces a similar effect, due to the movement of the armature in the circuit-closer.

Thus the presence both of mine and of circuit-closer, in a presumably efficient condition, are indicated.

The mine can, of course, be fired by the application of a positive current of sufficient strength at any time, and when it is desired to fire it automatically on a vessel striking the circuit-closer, a constant electric potential is kept on the system insufficient to actuate the armatures in either mine or circuit-closer, but sufficient when the resistance of the circuit to "earth" in the circuit-closer is greatly reduced to produce a current that not only attracts the mine armature, but also actuates certain apparatus on shore which automatically switches in the firing current and explodes the mine. The sketch on Fig. 57 shows this arrangement as described, but the instrument is provided with several terminals, and semi-permanent connections which can be easily altered, so that a large number of permutations and combinations are possible with two or more of these instruments, connected up so as to work differently with positive and negative currents.

Armstrong's Apparatus.

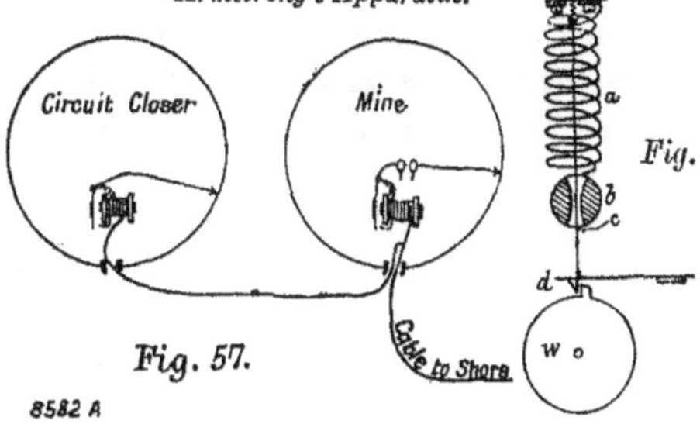

Fig. 57.

Fig. 58.

The ingenuity of the novice at submarine mining is, therefore, frequently directed towards the discovery of some new method of connecting up these instruments.

They have not been adopted by the Royal Navy, which service aims at the greatest simplicity in all arrangements connected with sea mining. For the same reason automatic signalling arrangements at the firing station are also omitted. The difficulty lies in obtaining high efficiency with great simplicity. This will be aimed at in the gear described on these pages. We shall thereby also steer clear of the apparatus adopted into our own service, which certainly is not remarkable for simplicity, however great may be its efficacy as claimed by those who have elaborated it.

Circuit-Closers.—A circuit-closer is a device for bridging a gap in an electric circuit when a vessel strikes the buoyant body containing the apparatus. The earlier forms were intricate, costly, and inefficient. For instance, the Austrian pattern exhibited at the Paris Exhibition of 1867 had nine projecting arms, each with a spiral spring, each with a water-tight joint, and all or any of them actuating a central ratchet wheel on a vessel striking the case. A partial revolution of the wheel produced the desired electric contact, and if the mine was not fired, the wheel and plunger or plungers returned to their normal positions.

A circuit-closer designed by Professor Abel, about same date, was a great improvement. The radial arms were replaced by a disc on the top of the case, and slightly larger in diameter. The disc was connected to the apparatus by a flexible and water-tight collar. When a vessel struck the edge of the disc it effected an electric contact in the apparatus, in a manner that can be readily imagined. The Abel circuit-closer is a reliable apparatus and possesses the advantage of insensitivity to signalling by the explosion of countermines in its vicinity. The pressures produced by such explosions might damage the flexible joint, but this could be guarded against without much difficulty. It is quite possible that some modification of the Abel circuit-closer may again be applied to submarine mining.

The circuit-closer which has found most favour, however, depends in its action upon the inertia of a small movable body placed inside the buoyant case. Such an apparatus, designed by Quarter-Master Sergeant Mathieson, R.E., was introduced at Chatham soon after the one just mentioned. It consisted of a lead ball on a steel spindle, the inertia of the ball causing the spindle to bend when the case was struck by a vessel, the flexure of the spindle producing the desired electric contact by means of a ring carried against suitable springs fixed radially around it. This circuit-closer was adopted into our service and used for many years.

In July, 1873, Mr. Mathieson patented certain improvements. After describing the above apparatus in his specification, he notes an important defect as follows: "This vibrating rod has hitherto consisted of a straight rod of steel, which, if not tempered to exactly the proper degree, will, when set violently vibrating by a passing vessel, either snap in two by being too hard a temper, or bend a little and take a permanent set if of too soft a temper, and thereby throw the adjustment out of order and render the apparatus useless." "To remedy these inconveniences I use instead of the straight steel rod a long length of stout wire coiled in the form of a helix, on which is fixed a short spindle with the weight on top. . . ." These apparatus in their turn were adopted into our service and a large number purchased. They are still serviceable and efficient.

The next improvement was designed by the author in December, 1876, and is shown on Fig. 58. *a* is a coiled spring of sufficient power to hold the small ball *b* in one position unless the apparatus receive a shock ; *c* is a silk cord connected with an adjusting screw *s* at one end, and with the spring detent *d* at the other. If a vessel strike the buoy containing this apparatus, the ball *b* is thrown sideways against the cord, and this pulls *d* and releases the wheel *w*, which is actuated by clockwork and makes a complete revolution slowly, during which period of time the cable is connected through the fuzes to "earth," and the mine can be fired if desired. This mechanical retardation gives time to the operators at the firing station to discover whether the circuit-closer has been operated by the shock of a countermine or by the blow from the vessel of a foe. From 300 to 400 contacts were obtained, when the clockwork ran down. This was its defect, for no record could easily be obtained showing the number of contacts expended. The apparatus was improved by Major (now Lieut.-Col.) R. Y. Armstrong, R.E., who substituted a small polarised electro-magnet for the clockwork, and arranged for the armature, normally out of the magnetic field, to be drawn up into same by the cord *c* when the circuit-closer is struck. There it remains until the mine is fired or until a suitable releasing current is sent through the electro-magnet, causing the armature to spring back into its normal position.

The scientific instrument makers of the School of Military Engineering have worked unremittingly for several years upon this germ, and have produced a complex instrument more suitable for the lecture table than for active service.

In December, 1881, the author drew up the following description of a *simple* apparatus which he believed, and still believes, is sufficient for all practical purposes. It was never forwarded officially, other work interfering, and was put aside until now. In Fig. 59, M M is a permanent horseshoe magnet, containing the ball-and-string apparatus already described. To the poles N. S. of the magnet are secured the cores of two small low-resistance electro-magnets C C, one end of the coil wire being connected to "line," and the other to a contact stud *b*. The armature A is secured by a spring to the fixed insulated point P, whence an insulated wire is carried through the fuzes to "earth." The other end of the armature spring carries a contact stud *a* which engages with *b*, when the armature is attracted to *n s* the poles of the electro-magnet, which are fitted with small ivory distance pegs, preventing absolute contact between the armature and the cores of the electro-

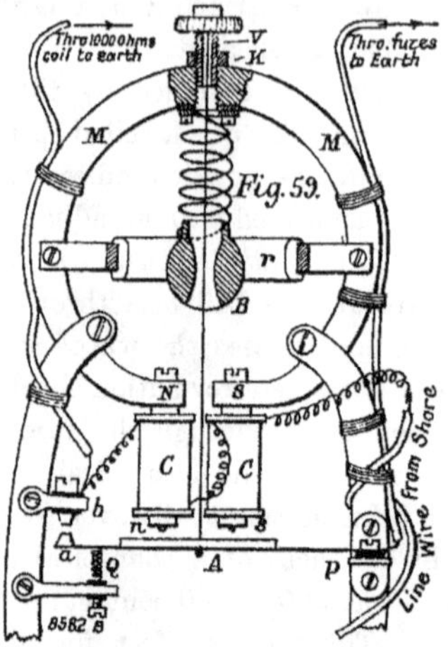

magnets, and thus avoiding magnetic adhesion. A set spring Q adjusts the strength of the armature spring. The magnets are shaped like those in an Ader's telephone. The top portion is perforated and tapped to carry the screw, and for regulating the tension of the pull cord, and a small nut K clamps same. An india-rubber ring *r* tied to a metal ring prevents the ball B from oscillating too violently. The various portions of the apparatus are clamped to, or carried by, strong brass standards, which are secured to a metal base.

When employed as a detached circuit-closer for a large mine below it, the stud *b* is connected to "earth" through an interposed resistance of about 1000 ohms, and in all cases P is connected to "earth" through the fuzes.

Firing by Observation.—The apparatus acts as follows. The coils C C are wound so that a negative current from shore increases the normal polarity of the soft iron cores, consequently when the negative pole of a firing battery is connected with "line," a current passes through the coils C C and the 1000 ohms resistance to "earth," causing the armature A to be attracted to the electro-magnet, and thereby shunting a current through the fuzes to "earth," the resistance on the fuze circuit being low enough to cause the mine to be exploded.

Firing by Contact.—On the other hand, if it be desired to fire by contact the negative pole of a weak but constant battery (a few Daniell's cells) is connected to "line," and when the circuit-closer is struck, the armature is pulled up mechanically and retained in that position magnetically. The signalling battery gives a signal on shore and the firing current can now be switched to line or not as desired, the mine struck being indicated at the firing station by a deflection on a galvanometer, and by causing an electric bell to be rung in a manner to be described hereafter, when the arrangements on shore are examined. Only one mine with a detached circuit-closer arranged in this manner can be put on one core.

Electro-Contact Mines.—When, however, the apparatus is employed in electro-contact mines, pure and simple, the "earth" wire from *b* through a 1000 ohms coil is omitted. Several mines, say six or seven, can then be connected with one core or single cable, either in a string, one after the other, or on fork (see Figs. 60 and 61).

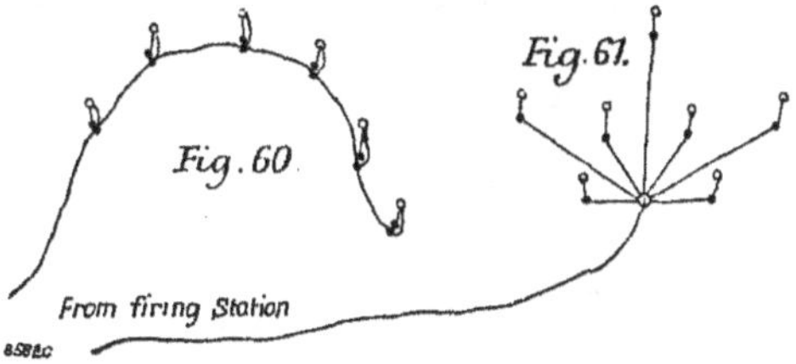

In the former case connecting boxes, as shown in Fig. 50 (see page 106), are used; in the latter a junction box, about to be described.

When any mine on a group is struck, the weak current battery in the firing station holds up the armature and deflects the galvanometer connected with the core leading to that group. The firing battery can then be connected to the core or not as desired. If it be not connected a positive current from the weak current battery releases the armature, by opposing the polarity induced in the electro-magnet cores by the permanent magnet, and brings the circuit-closing apparatus back to its normal condition.

This circuit-closer can be used in connection with automatic firing apparatus on shore, so that a firing battery common to a number of groups of mines is automatically switched to the core leading to the mine which is struck; but this leads to complications and difficulties, and a non-automatic system will probably be found to work more satisfactorily and in a simpler manner. Moreover, automatic firing is almost put out of court by countermining.

So much attention has been devoted to the systematic attack of mine fields in this manner, that some form of retardation in the circuit-closer is necessary, because these instruments, whether they can retard or not, generally signal at distances considerably greater than those at which their cases would be damaged. The operator on shore should therefore be able to fire a mine after the first signal, and when he has become assured that such signal is not caused by a countermine. Moreover, vessels, even so long ago as the American War of Secession, endeavoured to procure immunity from contact mines by submerged strikers rigged out in front of their bows to produce premature explosions.

But it is not absolutely necessary that the retardation should be produced by an electrical combination. It will probably be found that the arrangements, especially those in the firing station, will be simplified by the employment of a circuit-closer with a mechanical retardation, and an apparatus of this description has recently been designed by the author.

It consists of his ball-and-string arrangement, placed in a cylindrical chamber about 3½ in. diameter and high, the string actuating a rod R, Fig. 62, which carries a spring and contact maker C, which when

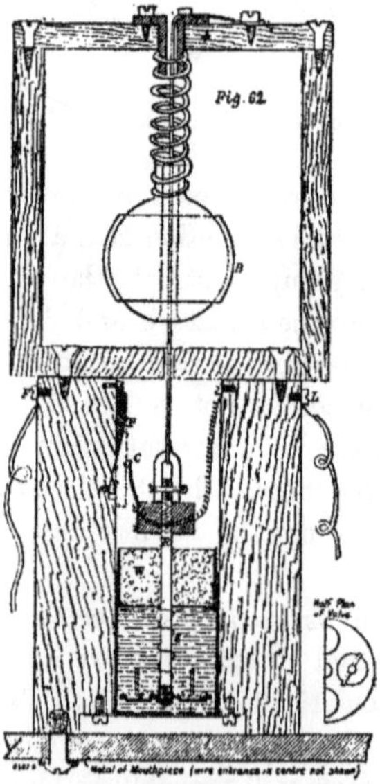

drawn upwards slides upon a platinised surface P, in connection with the terminal F, from which is led the wire to the fuzes. C is connected to the line wire at L, and the contact is prolonged far beyond the period occupied by the oscillations of the ball B, by the following device: The lower end of the rod R carries a piston, working in a small cylinder full of glycerine or other suitable liquid, the packing being a chamber full of cotton wool W, screwed into the top of the cylinder. The piston is kept in its normal position, and any slack taken out of the string by a small helical spring S, working on the rod R, and abutting against the bottom of the chamber W. The piston is fitted with four or more valves, each opening downwards and kept normally closed by a small spiral spring, acting on each valve spindle and abutting against a small crooked crossbar as shown in Fig. 62.

When the buoy or case carrying this circuit-closer is struck, the ball is thrown to one side by its inertia, the string is pulled through the guide hole, the valves open, and the piston rises, giving contact at C. The valves now close, and the glycerine has to leak past the piston while the spiral spring S gradually pushes it downwards. While this is going on the operator on shore can fire the mine if desired. The duration of the retardation can be adjusted to any required period by the space allowed between the piston and cylinder, and in settling this it will be useful to remember that a speed of 6 knots an hour is equivalent to 10 ft. per second, and that a vessel 300 ft. long would, at this speed, take 30 seconds to pass any given point. Even at 18 knot speed, 10 seconds would be occupied, thus giving an operator plenty of time to act with deliberation and discretion

Mercurial Contact Circuit-Closer.—Circuit-closers in which contact is made by the movement of mercury due to its inertia when the mine is struck, have been successfully applied. The idea was first brought to the notice of our Government in 1874 by Captain C. A. McEvoy, who has also invented several other circuit-closers on the inertia principle. They form a simple apparatus for mines which are not required to remain down a long time. When stores are improvised, the mercurial form of circuit-closer can be recommended; but it is most difficult to prevent a scum of oxide forming on the mercury, which may be thrown upon and then adheres to surfaces, thereby permanently bridging the electrical break in the circuit.

The apparatus may be made as follows: Cut $2\frac{1}{2}$ in. from a $\frac{1}{2}$ in. iron pipe; close one end with a metal plug; thread the other end internally; screw a wooden cork into it; bore a hole centrally in the cork; cut $1\frac{3}{4}$ in. from a No. 11 B.W.G. wire; thread it lightly end to end; fill the pipe one-third full with pure mercury; screw the wire through the cork until the end of wire is about its own diameter from the surface of the mercury; add a screw terminal on that portion of the wire projecting above the cork and the circuit-closer is made.

Contrivances have been designed to protect mercurial, as well as other forms of circuit-closers acting by the inertia principle, from the shocks of countermines, which are said to act upon buoyant bodies in a vertical direction. But, assuming this to be the case, it is evident that the contrivances must fail to act in the desired manner when the circuit-closers are out of the vertical themselves, the cases containing them being tilted by tidal currents or otherwise. It is consequently preferable to protect the mines from self-destruction caused by countermining, in some manner that will act efficiently under all conditions.

It should be noted that a mercurial circuit-closer does not and cannot be made to retard either mechanically or electrically. On the whole, therefore, it must be regarded as decidedly inferior to those patterns examples of which have been previously described in these pages.

Wire Entrances to Cases.—The electric wires can be carried into the case of a circuit-closer buoy or mine by means of an arrangement very similar to that already depicted in Fig. 37, page 74, and which it is not necessary to describe again.

Disconnecting Arrangements for Electro-Contact Mines.—The disconnecting arrangements, already referred to on page 116, should be as simple as possible.

Each branch cable to an electro-contact mine should pass through a disconnecting fuze placed in a water-tight case in the junction box. When a group of these mines gets out of order it is necessary to raise the junction box and discover the fault, which may exist either in the core leading to the group junction box, or in one of the branch cables or mines. Having raised the box, facilities should exist for testing each of these cores in succession, and this testing is generally done most conveniently from the firing station; signals, electric or visual, passing between the operator on shore and the party in the junction box boat.

When the mines are connected up on the fork system, it is necessary to provide a disconnector for each branch cable interposed between the cable and the mine. This can be done by placing a disconnector in a specially large connecting box on the branch cable and close to the sinker of each mine. But it is far better to place all the disconnectors in the junction box. Various devices have been tried for combining the several disconnectors into one apparatus forming part of a special junction box, and for combining a commutator therewith which can be plugged or unplugged when the box is lifted and opened, so that tests may be taken from shore to each mine. Also an apparatus has been proposed, and I believe patented, for electrically actuating such a commutator from the shore, and thus testing daily each electro-contact mine in turn. These shore tests will not rectify a fault, and a simple resistance test for the whole group indicates with sufficient clearness whether a group box should be raised and a group examined. The more complicated tests tell us very little more, and the additional apparatus are liable to derangement. The first step to be taken for the recovery of a faulty mine or cable is the raising of the junction box, and no important economy of time is effected by localising the fault beforehand. On the whole it is better to have a separate case for each disconnecting fuze, and the arrangement shown on Fig. 63 (now designed) would act efficiently.

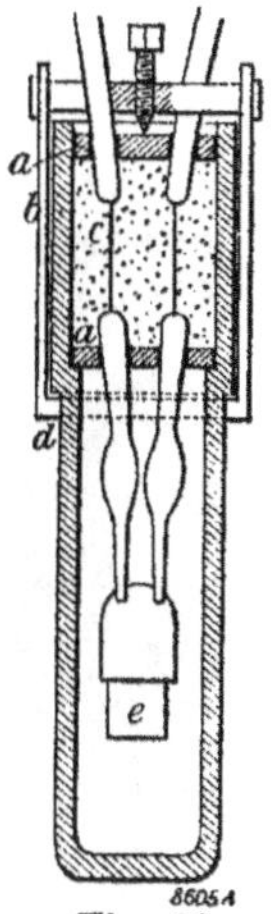

Fig. 63.

The upper cylinder, containing the india-rubber plug *c*, should be turned out internally, but the rest of the case may remain rough cast. The gear consists of an ordinary screw bolt working through a cross-bar, supported by two side links *b* from an encircling ring *d* slipped over the case as far as its smaller diameter is carried. The screw bolt squeezes the india-rubber plug *c* between the two iron plates, thus forming a water-tight joint for the wire entrance to the chamber containing the disconnecting fuze *e*. These apparatus being carefully made up on shore, should never require to be opened on the mine field, where one leg is connected by an insulated joint to the branch wire leading to a mine, and the other leg to the core of the single group cable. But this gives no facility for rapid testing. For this purpose a multiple connector of some sort should be interposed between the disconnectors and the group cable, and it should be so arranged that it can be easily opened, without disturbing any electrical joint, and each branch cable or the group cable tested. The apparatus shown in Fig. 64 is now designed by the writer, and will serve as an example of what is required. The cylindrical case is open at each end and has an internal shoulder near one end against which a wooden disc *f* abuts. A water-tight joint, similar to one just described, is formed by the rubber plug *g*, the iron plate *h*, and the bolt and crossbar *i*. The wires from the group cables and the wire from the main core are led through and connected to brass terminal screws *e e* secured in the top of the wooden disc. This portion of the apparatus need not be opened on the mine field. The upper end of the cylinder is closed by a rubber ring *c*, a cover *b*, and a crossbar with screw bolt *a*. The brass terminals are provided with additional screws whereby a thin piece of sheet brass is connected to all from the central terminal, which is slightly higher than those around it. Consequently, when the binding screws that secure the sheet brass to the latter are released, the brass will spring up and be out of contact with them. Plugs are liable to be shaken out by

Fig. 64.

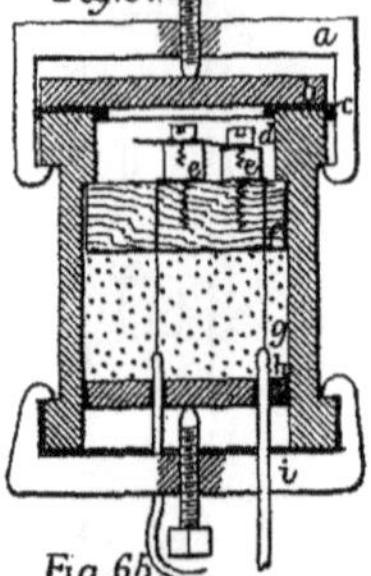

Fig. 65.

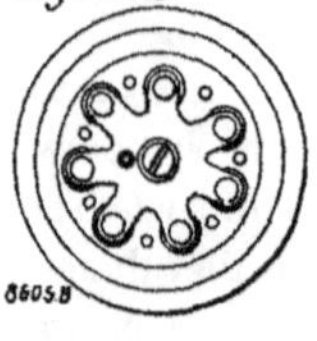

neighbouring explosions, and should not be used. An earth plate over side of boat can then be connected with the central terminal, and the core to firing station tested for resistance. If good, the earth plate wire is removed, and each branch cable terminal screwed down in turn, and the several resistances tested from shore. As soon as the faulty cable is found, it should be disconnected from the junction box, under-run, and the mine picked up, taken ashore, and the defect discovered and made good.

The general arrangement in the group junction box is shown in Fig. 66 (the section is similar to that shown in Fig. 51, page 106, and the electrical joints between the seven single disconnectors and the multiple connector may be made on shore, and if done carefully should never require to be touched again on the mine field.

The number of mines on one group cable depends upon the electrical system employed. When the simple arrangements advocated in these pages are adopted, each mine when perfect testing infinity, any number of mines can theoretically be connected in one group, but, as one faulty

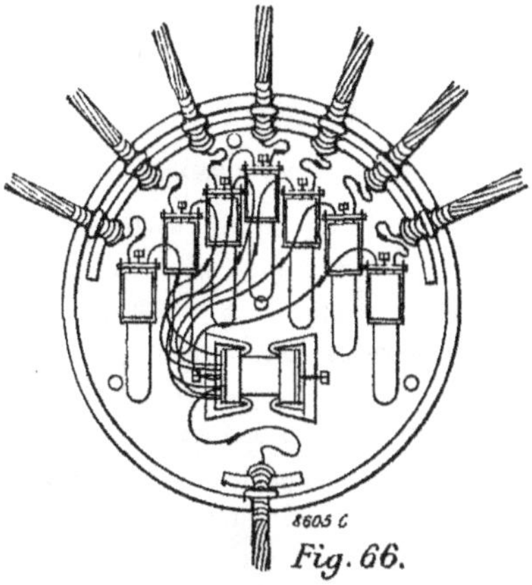

Fig. 66.

mine or cable destroys the efficiency of the group for the time the fault lasts, it is better to limit the number of mines to six or seven. In the English service, when Colonel Armstrong's testing apparatus is employed in each mine, the number in a group has to be still further restricted.

It will now be convenient to examine the electrical arrangements which are required on shore for controlling and firing the mines, &c., that have been described.

CHAPTER XI.

Electrical Arrangements on Shore.

Electro-Contact Mines.—Continuing with electro-contact mines, the sea arrangements for which have just been explained, the first and most important thing on shore is the firing battery.

It is a matter for serious consideration whether a small dynamo driven by hand should not be employed for the purpose, rather than a voltaic battery. For the arrangements hereinbefore recommended, as the group mines may perhaps be as far off as three sea miles, the external resistances may be 108 ohms for cable core, 2 ohms for two earths, 4 ohms for four fuzes, or a total of 114 ohms. The minimum firing current being 0.17 ampère, the current used should be 0.5 ampère (top page 116). Consequently the dynamo should, at the speed driven, be able to produce a current of 0.5 ampère through an external resistance of 114 ohms.

Following the beaten track, however, the best voltaic battery to use for such a purpose is that known as the Leclanché, with an electromotive force per cell of 1.45 volts, and the cell usually employed and specially manufactured for the purpose, has an internal resistance of about 0.3 ohm. But the fuze which it is now proposed to employ has a resistance of 1 ohm, and the current required being 0.5 ampère, such a fuze on short circuit would not require so large a cell. Thus, if the internal resistance of the cell were 1 ohm instead of 0.3, it would possess ample power and would give a current through a 1-ohm fuze on short circuit of $1.45 \div (1+1) = 0.725$ ampère. A Leclanché firing cell with 1 ohm resistance is, therefore, recommended when the mines are fired by fuzes as sensitive as those now employed by the Royal Engineers for land service.

With 114 ohms external resistance, if N be the number of cells required in battery,

$$0.5 = 1.45\ \mathrm{N} \div (\mathrm{N} + 114)\ .\ \therefore\ \mathrm{N} = 50 \text{ cells.}$$

But the mines are not usually so far off, and perhaps one mile may be taken as an average distance in most harbours.

The external resistance then = 42 ohms, and the equation becomes

$$0.5 = 1.45\ \mathrm{N} \div (\mathrm{N} \div 42)\ .\ \therefore\ \mathrm{N} = 19 \text{ cells.}$$

For Distances over 1000 *Yards if D be the Distance in Yards,*

$$N = D \div 1000 \text{ (nearly)}.$$

"The Silvertown firing battery," Leclanché, "is put up in stout boxes containing ten cells coupled permanently in series with two terminals outside. Each cell is sealed, and contains all the parts needful for action except water, which is to be introduced through two holes in the top introduced for the purpose. The cells are made of ebonite. The zinc plate . . . is a cylinder . . . surrounded with a packing of sal-ammoniac in powder, enough being inserted to more than saturate the charge of water . . . The negative element in its present agglomerate form consists of a central carbon hexagon grooved on each side to fit a cylinder of compressed peroxide of manganese and carbon, 6 in. long and .9 in. in diameter. The whole are wrapped with a strip of burlap held in place by a couple of rubber bands. Each cell is 4 in. in diameter and 7¼ in. high, and should receive about eight fluid ounces of water when the battery is removed from store for use in service." . . .

"The great fault of the arrangement is the insertion of the powdered sal-ammoniac; but the sealing is also a defect. The salt contains sufficient moisture to slowly encrust the zinc with a coating of oxy-chloride crystals, which, being insoluble in the added water, increases the internal resistance much above its normal value. To remove these incrustations it is best to cut through the pitch covering, take out and wash the zincs in a strong mixture of muriatic acid and water, re-amalgamate them, and replace them. . . . The cells should never be resealed. Two bits of marline saturated in paraffin and packed" . . . on either side of the zinc "sufficiently prevent evaporation, and are far more convenient than the pitch cover. The cells, when required for use, should be charged with a saturated solution of sal-ammoniac, with a little of the salt added to supply consumption, the zincs being first re-amalgamated." . . . "A firing battery of forty of these cells was set up . . . and kept in active service for over six years" (Abbot). As before stated, the internal resistance of the cell above described is .3 ohm, and for the fuzes now recommended a resistance of 1 ohm is permissible. A cell of smaller dimensions, or one of simpler construction, may therefore be employed, and the cell recently patented by M. Leclanché will possibly be found to answer well. This arrangement is illustrated by Figs. 67 and 68, and consists of an outer glass jar A containing an exciting solution of chloride of ammonium, or an acid or alkali, in which is immersed a central cylinder D of zinc. The positive electrode is formed by an outer hollow cylinder B of special depolarising composition. The part of this cylinder which is above the solution is paraffined, and has a ring E of lead or other metal firmly secured to it.

The cylinder B is also provided with holes I to allow free passage to the exciting liquid. The cylinder D is kept in place by a stopper F of wood or ebonite. A caoutchouc ring G prevents evaporation of the liquid. The cylinder B is composed of a mixture of peroxide of manganese, graphite, pitch, and sulphur, moistened with water, and pressed into shape while cold and then baked. The operation of baking induces partial volatilisation and vulcanisation of the composition, which is thereby rendered porous and a good conductor of electricity.

Electro-contact mines not fitted with testing apparatus can, if they are all in good order, be connected direct to the firing battery. But there are several objections to such a proceeding: 1. It is peculiarly vulnerable to attack by countermines. 2. A number of mines are very rarely "all in good order." 3. Boats and steamers connected with the defence may accidentally come in contact with the mines. 4. The mines may signal by wave action. 5. Some of the mines may drag their moorings, and the explosion of one mine may then cause others to signal and be exploded.

Fig. 67.

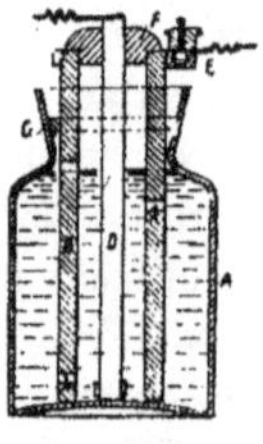

Fig. 68.

Fig. 69.

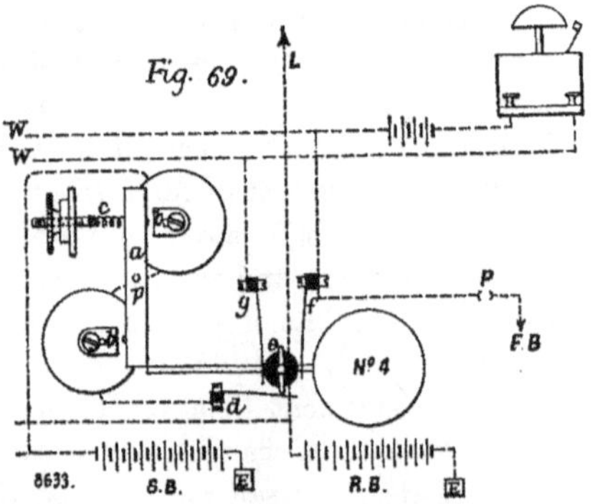

For these and other reasons the firing battery should not be connected direct to line, and it becomes necessary to devise some simple arrangement of apparatus for employment in the firing station, so that the mines may be under control, and act when and how desired, and then and thus only. If, in addition, the apparatus so employed give a record of the number of mines fired in each group, so much the better.

The plan usually pursued is to place a small electric current continuously on "line," and when the circuit-closer in or above the sea mine is actuated by a passing vessel, or by a countermine, this current is increased by a decrease of circuit resistance, so that an electro-magnet on shore moves an armature and a dropping shutter, which automatically closes the firing circuit, rings a bell, &c.

Such an apparatus is shown in Fig. 69, which was one of Mathieson's first designs for a shutter apparatus, except that I now add a bell circuit and spring *g* for same. The more intricate apparatus since elaborated for the English service cannot beat this simple first form.

Plugs, not shown on Fig. 69, should be provided for disconnecting the batteries S B and R B, as also the leading wire L from each shutter axis.

An armature *a* pivots on *p* between the two horns *b b* of an electro-magnet, small ivory studs preventing actual contact between them. The lever of a weighted shutter (No. 4) engages the lower end of the armature, so that when the armature is attracted by the electro-magnet the shutter falls. This occurs when the resistance of line is decreased by a contact made at the circuit-closer in one of the mines of the groups connected to L. The axis of the shutter is insulated and connected to L, and the metal crossbar *e* is normally in contact with the spring *d*. As soon as the shutter falls *d* is automatically disconnected, and the firing battery F B is connected direct to L through the spring *f* if the firing plug P has been inserted. In general a small bell is struck mechanically by the falling shutter. But I prefer to employ a ring bell on a local circuit arranged as shown in Fig. 69. This arrangement does not affect the firing battery, although the firing battery spring is used for it. The wires W W lead to the springs *g f* of the other shutters. The signalling battery S B can be common to the seven apparatus in one set. The firing battery F B and the releasing battery R B can be common to a number of sets. A releasing battery is of course only required when a circuit-closer is used that can retard. Mathieson's shutter apparatus was designed about the year 1870, and has been employed in our service in a modified form ever since. An apparatus was patented by Captain McEvoy in 1884, by which similar actions were produced by a shutter falling between guides; but the pendulum action is preferable, as it is less likely to be affected by the concussions due to the firing of heavy artillery, and the axis of the pendulum motion affords a more reliable method of changing the connections.

If the circuit-closing arrangement in the sea be made to retard either magnetically or mechanically, the firing battery can be, and generally is, plugged after the shutter falls, and in this manner the self-destruction of sea mines by the concussions caused by countermines may be obviated, means being provided to acquaint the operator with the operations that are proceeding in the water.

The current required to actuate such an automatic system cannot be much less than 0.15 ampère (American system, Abbot), and the fuzes employed should therefore not fire with less than $\sqrt{10(0.15)^2}$ or 0.47 ampère. The firing current to insure simultaneous ignition of fuzes in mine and disconnector should, therefore, be 1 ampère. A shutter apparatus may be accidentally actuated by the concussions produced by the discharge of large guns in its vicinity, unless care be taken to guard against it. Again, the development of the attack of sea mines by countermining almost prohibits the use of a purely automatic method of firing, and if we are never to use the shutter apparatus in this manner there is nothing to recommend it in preference to simpler arrangements that depend upon the vigilance of an operator. In designing such an arrangement it is of the utmost importance to remember that the number of highly trained electricians available in time of war may be, and probably will be, limited. If, therefore, the arrangments can be worked by men of ordinary intelligence, by following some simple and clear instructions, a great advantage will be gained. Many arrangements, much simpler than those now in vogue, can no doubt be elaborated, and the following is given as an example. It has been designed by the writer as he penned these pages, and appears to be a simple solution to the problem. Advantage has been taken of the theory of the simultaneous ignitions of low-resistance fuzes already explained.

About two years ago a naval officer in the Vernon Torpedo School brought to my notice the advantageous use of what he termed a protecting fuze in the firing station. His idea was to use one such fuze and replace it when expended. I now propose to enlarge upon this idea, and to place a number of such fuzes systematically on the firing bar, one for each mine, and to plug each in rotation in every group of mines.

The firing of each fuze will then not only "protect" the remainder of the mines in that group from premature and undesired explosion, but will also indicate that a mine has fired, and remain as a lasting indication thereof.

Let us assume that the mine fuzes have a firing current sensitivity of about 0.17 ampère, and that they are placed out of circuit until the circuit-closer is actuated, as described in the recent chapter on circuit-closers. Also that the latter are provided with a magnetic retardation releasable at will by a suitable current from the firing station, or with a mechanical retardation lasting for several seconds before the circuit is again opened at the circuit-closer. As many as seven mines, and even more, can then be placed on each cable core (see plan of mine field on sketch, Fig. 70).*

Each core is led through a disconnecting fuze and a multiple connector in the group junction-box (Roman figures on plan) to the firing

* In this figure the scale above X Y is about 1 in 1000, and below X Y it is about 1 in 8.

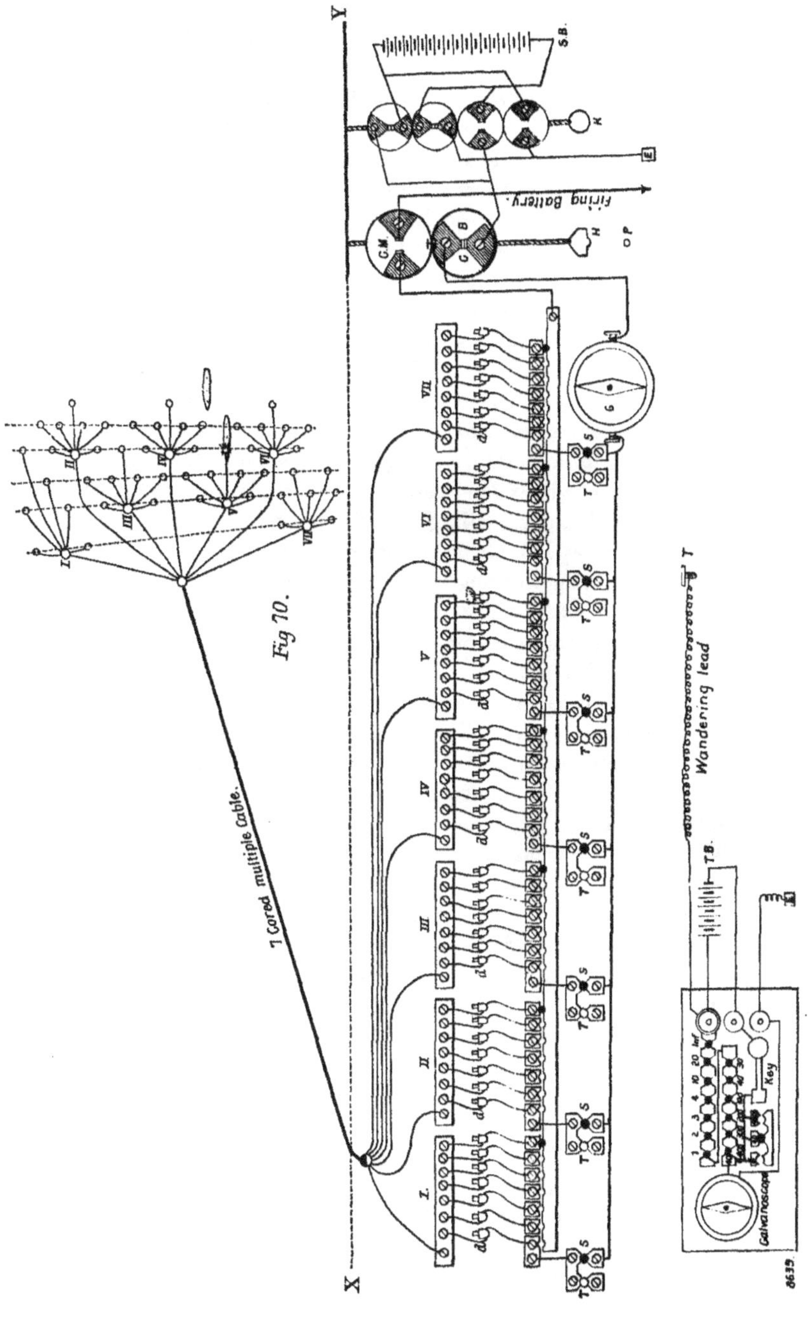
X
Y
7 Cored multiple Cable.
S.B.
Firing Battery.
C.M.
Wandering lead
T.B.
Key
Galvanoscope

Fig 10.

station, where the path is split, the road to the firing battery passing through another disconnecting fuze to the firing bar, and thence through one of my patent (pull) contact-makers C M to the negative pole of the firing battery F B and "earth," C M being normally open, and the other route passing through a disconnecting fuze to the signal battery bar, and thence through a galvanometer and one of my patent (pull) circuit-breakers C B to the signal battery S B and "earth." C M and C B are actuated simultaneously by a pull on the cord from the handle H. By pulling this handle and securing it on the peg P, the arrangement becomes automatic. With magnetic retardation requiring a positive releasing current the employment of a separate releasing battery can be avoided by sending a reversed current from the signalling battery to line. This can readily be done as indicated on sketch by the employment of two of my contact-makers and two of my contact-breakers actuated by one pull cord and handle K, the connections being made as shown in Fig. 70. The tests for resistance of each may be taken daily (and perhaps oftener), group by group, without interfering with the signalling and firing arrangement of the other groups if the wires and plugging plates be arranged as indicated, the plug S and the plug to firing bar being removed from any group which is to be tested, and the wandering lead T inserted in plug-hole T of that group. The apparatus recommended for this test, and shown on the plan, consists of a battery, galvanometer, key, and a set of coils with bridge which is capable of testing resistances from $\frac{6}{10}$ ohm to 11,000 ohms, an ample range for all sea mine purposes. The whole, including battery, is contained in a box 9 in. by 6 in. in plan, and one such apparatus will probably be enough for one firing station. (Makers—Elliott Brothers, London.)

The firing battery may be common to a number of similar arrangements in one firing station. Each firing station should be provided with an electric bell under the control of an observer placed so as to command a good view of the mined waters and channels of approach, probably at one of the stations for observation firing. He would then control the firing of both the observation and electro-contact mines.

This observer may advantageously be connected with submerged telephones so that the explosion of countermines may be detected by him. If for this, or other cause, he considers that the electro-contact mines should not be fired, he rings the caution bell in the firing station. If possible he should be in telegraphic communication with the officer in command of the picket boats. When the caution bell is ringing in the firing station, the handle H must not be touched; and if G deflect at this time a quick pull on the handle K should bring it back to normal. If not, this pull on K should be repeated. When the caution bell is not ringing the handle H must be pulled when G deflects. This should fire a mine, also one of the indicating and protecting fuzes, thus preventing any other mine in the group being fired, and if the groups be separated a little more than shown on diagram, no fear need be entertained that a mine in one group will cause one in another group to explode. As soon as a fuze is fired, the handle H should be released, and if G still deflect a pull on K should free it. Another fuze should then be plugged to the firing bar, for that group.

These operations are simplified when mechanical retardation is used in the sea circuit-closers, the releasing current being omitted, and the pull handle K, &c.

The fuzes when fired should not be removed. They then form a record of the mines expended.

Faults.—Should a group test low, a faulty branch can sometimes be disconnected by the firing battery, the fault then being beyond a group junction-box disconnector. A fuze should fire at the firing station, and the deflection on G go to normal when H is released. Another fuze can then be plugged to the firing bar, and the remaining mines of the group become effective. This drastic method should only be resorted to by command of an officer, who should order same only when it is more important to have, say, six mines effective at once than seven a few hours later. If repair be decided upon, the faulty group must be disconnected from the system by unplugging the fuze to firing bar and removing plug S of the group. The group junction-box must then be raised, the multiple connector opened, the wandering lead T plugged in the plug-hole T, of group at firing station, and each core tested as before explained. The fault should then be rectified by laying a new mine or by other means, and the group efficiency recovered.

The electromotive force of the voltaic batteries employed in this arrangement being low, it is not necessary to use ebonite for insulating purposes. The plugging brasses can therefore be secured to a kiln-dried hard wood backing protected from damp by hard varnish. Teak is probably the best wood to employ, certain experiments instituted by the United Telephone Company having given the following comparative results :

TABLE XXXII.—ELECTRICAL RESISTANCES. WOOD OF SORTS.

Mahogany, resistance, comparative, along fibre ...				..	40
Pine	,,	,,	,,	... ...	214
Rosewood	,,	,,	,,	... ...	291
Beech	,,	,,	,,	... ...	397
Oak	,,	,,	,,	... ...	478
Teak	,,	,,	,,	... ...	734

Resistances across fibre are from 50 to 100 per cent. greater.

The firing station must, of course, be kept dry by means of artificial heat.

Instructions for Operator.

Caution Bell Ringing.—G deflects. Pull K. Deflection should vanish. If not, repeat.

Caution Bell not Ringing.—G deflects; pull H. A fuze should fire. If not, repeat. G should now cease to deflect. If not, pull K, and this will occur. Plug another fuze to firing bar.

With mechanical retardation the pull cord and handle K are omitted, and the instructions are still simpler, thus:

Caution Bell Ringing.—Do nothing, whether G deflects or not.

Caution Bell not Ringing.—G deflects; pull H; fuze should fire. If not, repeat. G ceases to deflect; plug another fuze to firing bar.

The signalling battery employed should be capable of working continuously on a somewhat leaky line. A single fluid gravity Daniell cell is probably the best suited for such conditions. To avoid any possibility of accidental explosions the resistance of each cell should be such that the current on short circuit through one of the fuzes employed does not exceed 0.054 ampère (see page 117). Hence, the electromotive force of a cell being 1 volt, and the fuze resistance (cold) 1.6 ohms, if x be the liquid resistance, we have $0.054 = 1 \div (1.6 + x)$, from which $x = 17$ ohms. The size and arrangement of each cell should therefore be such that it possesses this amount of resistance.

Looking back to the circuit-closer recommended for magnetic retardation (page 124), it will be seen that the signalling current is required to perform no work except moving a galvanometer when an electro-contact mine signals. Retardation is effected by the induced magnetism from the permanent magnet holding up the armature. But the releasing current has to perform work, viz., to neutralise the said magnetism in order to produce the fall of the armature. The coil in the circuit-closer, the strength of the magnet, and the size of the ivory stops, must, therefore, be so adjusted that a current of, say, 0.054 ampère from the signalling battery reversed will act as desired. Assuming that this adjustment has been secured by the instrument maker, the number of cells required can be calculated. Let the line fuzes, earths, and circuit-closer coil amount to a total external resistance of 80 ohms, and six cells will be the number required in the signalling battery.

The battery employed for the resistance tests can be a high-resistance Leclanché.

Large Mines provided with Detached Circuit-Closers can be arranged in a precisely similar manner if they are not to be fired by observation as well, but such mines must be spaced in group, so that they will not damage one another, or their detached circuit-closers (see page 87).

If, therefore, the groups be spaced so that the explosion of a mine will not signal those in another group, the system now recommended enables us to place such mines in group much nearer to each other than is possible by the methods of shutter apparatus usually employed. Mines with detached circuit-closers, arranged for purely contact firing, then become a formidable defence (page 78). Hitherto it has been necessary to space them so far apart that a great expenditure of cable became necessary in connection with their employment. For important harbours the flanks of the narrow navigating channels may therefore be mined in future on the plan suggested. The experiments against H.M.S. Resistance, and the provision of numerous water-tight compartments in modern war vessels, favour the employment of large charges that rack and shake a ship from stem to stern rather than small charges, which are far more local in effect.

The employment of apparatus whereby large charges can be fired either by a detached circuit-closer or by observation has already been alluded to on page 124. Arrangements of the kind, giving the two methods of ignition, are old, and have fallen into disuse for some years. They produce complications in the firing arrangements, and, as a number of observation mines must be used in the channels kept open for traffic, the trained observers available at any one port are likely to be fully employed in working them, and the system cannot be recommended.

Observation Mines.—Firing by Single Observation.—Large sea mines, situated near to an observing station, are sometimes fired by one observer when a vessel is between marking buoys that indicate the position of the mines. When this plan is resorted to, several mines (three, four, or five) are usually connected up in one line and fired simultaneously, the marking buoys being moored near to the extremities of the lines, which are placed across the channel. Reverting to Fig. 39, page 79, taking the beam of a vessel at 60 ft., and the spacing between the mines at 60 ft. more, or a horizontal striking distance for each mine of 30 ft., it will be seen from Table XXIV., page 80, that the strike of each mine should be 39 ft. for 40 ft. depth of water at high tide, $42\frac{1}{2}$ ft. strike for 50 ft. depth, 52 ft. strike for 60 ft. depth, 60 ft. strike for 70 ft. depth, 68 ft. strike for 80 ft. depth, 76 ft. strike for 90 ft. depth, and 85 ft. strike for 100 ft. depth.

Charges Required at Various Depths.

Comparing these figures with the effective striking distances of the mines suggested for adoption on pages 83 to 86, it will be found that the small ground mine loaded with gun-cotton will not act efficiently in a line of mines spaced at 120 ft. intervals even at so small a depth as 40 ft. Dynamite can be used—not because it is stronger per pound, but because the case will hold a heavier charge, giving a strike of 45 ft. At 50 ft. depth the same mines with dynamite or stronger explosive may be used. If gun-cotton only be available the large ground mines must be used at depths of 40 ft. and 50 ft. At 60 ft., the small ground mine with gelatine dynamite or stronger explosive may be used. At 70 ft. depth the small ground mine with explosive gelatine, or the large ground mine with dynamite or stronger explosive, may be used. At 80 ft. depth the large ground mine with dynamite or stronger explosive may be used. At 90 ft. depth the large ground mine with gelatine dynamite, or stronger explosive may be used. At 100 ft. depth the large ground mine with explosive gelatine is required. Above 100 ft. depth either the large or the small buoyant mine may be used, and be loaded as desired, the submersion being regulated accordingly.

By this system, a line of two mines will cover 180 ft. of cross channel, three mines 240 ft., four mines 300 ft., and five mines 360 ft.

When it is desired to keep a still wider channel than 360 ft. open for traffic, and therefore clear of all E.C. mines or other obstructions, it can be done by providing a line of fairway buoys down the centre of the channel, and a line of boundary buoys on either side of the channel. One half channel can then be used for up and the other for down traffic. Lines of mines can be placed in each half channel, and if lines of, say, four mines each be used the total width will be 600 ft., which ought to be sufficient even for the Thames or Mersey. This is perhaps better than providing two independent and separate channels, each 300 ft. wide, because the traffic vessels could keep near to the fairway buoys, and thereby be certain not to damage any E.C. mines near the boundary lines.

The marking buoys should be distinguished by shape rather than colour, because colours cannot be seen well at night, whereas the electric light brings out the shape of any object with clear definition, especially if it be painted white or red.

One observer can look after three or four lines of mines, and perhaps as many as six lines, but this is putting too many eggs in one basket. The observer should have an assistant (out of fire) to take his place in the event of a casualty.

There are other methods of firing by single observation which require the observing station to possess a good command (in height) over the mined waters. The camera obscura is so used in the Austrian service; but it fails in bright moonlight, when the picture becomes so obscure that many objects seen clearly with the naked eye direct, cannot be distinguished upon it. Other arrangements for firing mines by a single observer placed some distance above the sea level have been made.

One of the first, an instrument designed by Major H. S. S. Watkin, R.A., was a development from his well-known depression range finder; but was complicated by springs, chains, pulleys, &c., in order to obtain the movement of an indicator over an adjoining chart, which represented the mined waters. Another, designed by the writer, had an arm with terminal pointer connected to the telescope axis by elliptical gearing, so as partially to counteract the reduction of scale due to perspective, and this pointer traversed a chart pasted on a nearly spherical surface fixed behind the instrument and observer. There was no correction for rise and fall of tide, and it consequently failed in most situations. Finally Major Watkins invented an excellent instrument of comparatively simple construction which has been adopted for employment in our service. Unfortunately this instrument cannot be described.

The theoretical considerations underlying the construction of all depression instruments that possess a correction for alteration in tidal level is shown on Fig. 71, where A C represents the telescope's

Fig. 71.

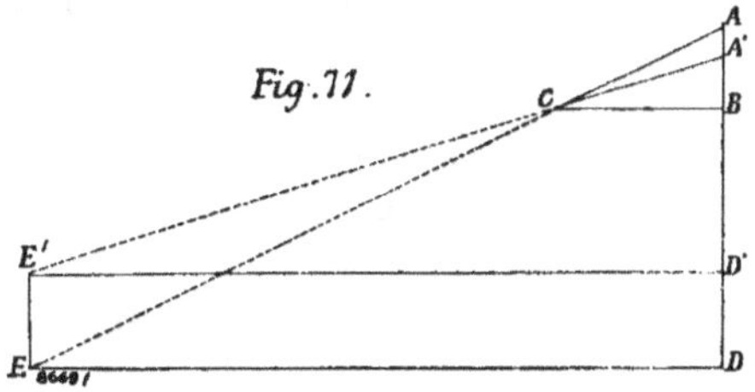

collimation, B C the horizontal scale employed, D E the water surface, A B the scaled height of telescope, A D its actual height above water level. If the latter be altered to D¹ from a tidal rise, or A D decreased to A D¹, then in order to obtain a tidal correction in the instrument, A B must be proportionally decreased to A¹ B, and if A¹ and E¹ be joined they will still pass through C, the length of C B still indicating the distance of the mine E from the vertical A D.

The inherent defect of all depression instruments, whether for finding range or for plotting the positions of objects on a chart, consists in the difficulty encountered in finding visually the water line of the object observed.

At night it is especially difficult to see the water line of a black hull moving slowly through water that also appears to be black; and if the vessel move fast, the bow wave and wave of depression behind it make it almost impossible to discover the true level of the water line.

It has been proposed to employ depression instruments with a command of 50 ft. or 60 ft., say, 20 yards.

With such a command an error of one vertical yard in taking the observation causes an error in range of 48 yards at 1000 yards if the vertical error be downwards, and of 52 yards if it be upwards. As the vessel itself is, say, 20 yards wide, the latter error would be reduced to 32 yards. It may be urged that the cut-water is the proper point to observe. But the bow wave would make such an observation unreliable. Approximate accuracy can only be obtained by using one of the crosshairs horizontally, then lining it with the water line on the vessel's side nearest to the observer, and thereby obtaining the mean or average line of flotation. Remembering that the large mines cannot conveniently give more than 30 ft. or 10 yards radius to the circle of observation, it appears evident that depression instruments, however perfect in themselves, should not be employed when the command obtainable does not exceed 60 ft. At what command can their employment be permitted?

Assuming that the instrument must be effective at night, and taking into consideration the undeniable fact that the vessels to be observed will probably be enveloped in smoke, that their water line will frequently be invisible, and will often have to be guessed at from short glimpses at the remainder of the hull, those who are unbiassed will agree that under such circumstances a vertical error proportional to 1 in 1000 of range is likely to be made by the coolest and most careful observer.

Let H be the height of instrument above water level which it is desired to find.

Let D be the horizontal distance of the mine M and r its horizontal distance from the ship's side S (see Fig. 72).

Fig. 72.

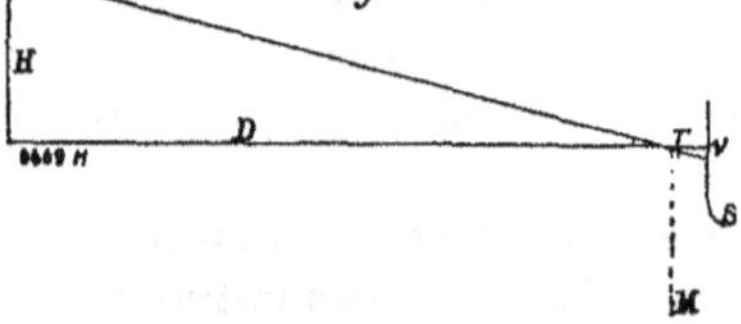

Then if v be the vertical error of observation, r becomes the horizontal error, and

$$H : D :: v : r.$$

But

$$v = D \div 1000.$$

Consequently

$$H = D^2 \div 1000\,r.$$

But r should not exceed the radius of horizontal effect of the mine. Consequently, the height of a depression instrument should not be less than the square of the distance in yards of the furthest mine, divided by 1000 times the horizontal radius of effect of the mine, which latter may be taken at 10 yards for a mine charged with 500 lb. of a high explosive. Thus:

When D = 500 yards	H = 25 yards = 75 feet
D = 1000 ,,	H = 100 ,, = 300 ,,
D = 1500 ,,	H = 225 ,, = 675 ,,

Depression instruments do not therefore possess sufficient accuracy for firing mines singly during the smoke of an engagement, and the high value claimed for this system of firing is certainly not deserved.

When the probable course of the vessel of a foe is perpendicular to the line of sight A E, Fig. 71, two or more mines can be placed on one core and be fired simultaneously, the point observed being the centre of such a group, and the mines being each moored on the line of sight. The spacing of such mines should follow the same rule as that for lines of mines between marking buoys, already given, viz., 120 ft. Consequently the width of channel covered by a pair of mines would be 180 ft., and r in above proportion becomes $180 \div 2 \times 3$ yards $= 30$ yards, instead of 10 yards, and the height of the depression instrument may be reduced accordingly. Thus:

When D = 500 yards	H = 25 feet
D = 1000 ,,	H = 100 ,,
D = 1500 ,,	H = 225 ,,

These figures indicate that mines in pairs should generally be employed in connection with single observation instruments working by depression.

The great defect in this system of firing mines, viz., the impossibility to observe the water line of a vessel when she is enveloped in dense smoke, is not shared by the system in which a mine is fired when it coincides with the intersection on plan of two separate lines of sight, because the mast of a vessel is seldom obscured by smoke. Instruments for firing mines by the intersection of lines lying on the horizontal plane are therefore much to be preferred. Nevertheless, the fact should be recorded, that adepts in this country do not, as a rule, share this opinion, and for several years firing by double observation has therefore received but little attention.

Firing by Double Observation.—Accuracy, that all-important element, can be absolutely assured when the base is not less than about one-third of D, the distance of the furthest mine. In other words, the angle of intersection should not be less than 20 deg. This can always be easily obtained by the horizontal intersection of two lines of sight, but two separate observers are required. Instruments have been designed by which quite a short base is used, and the two observers are therefore placed close to each other. It is quite impossible for such instruments to work with sufficient accuracy to fire distant mines at the desired moment. The objection to firing by cross intersection, or double observation, is the liability of the two men to observe different ships or different parts of the same ship. But the employment of the electric search light facilitates matters very considerably, a rule being followed that the two observers shall sight on that vessel which is illuminated by the ray of light, and stick to her as long as the light remains upon her. Also a rule should be made that the centre of the foremast shall be the point of observation; and the difficulty is reduced to a minimum.

The observers for double observation firing should be situated at some distance from the smoke of a battery, and would probably be more secure from machine-gun fire when placed at some considerable elevation above the mined waters.

Various instruments have been elaborated for firing by intersection. One of the first arrangements with which fairly accurate results were obtained up to ranges of 1000 yards, consisted in placing a number of small pickets on the circumference of a circle about 20 ft. radius, a central picket giving the position of the observer's eye. This acts well enough as an improvised arrangement, when the instruments for observation firing have been damaged.

If the space required be not available, the scale can be reduced by employing a flat board in the form of a segment of a circle, and placing small sights on the circumference. Also, if desired, the centre of the segment may form the outer sight, and the eye of the observer be placed at each of the inner sights in succession. By this plan the observations may be taken through an aperture not much larger than the loophole for a musket.

The mines to be fired by double observation are usually laid down in rows converging on the advance station B (see Fig. 73), which is therefore generally set back at some distance from the river bank. This is advantageous, for it gives more security to the observing station from attacking parties advancing by boats. The firing battery should be at A between A and B. Station A should be so placed that B can be seen. Should the cable between them become damaged, visual signals can then be used, and the battery "earthed" at A.

An advanced station is sometimes impracticable. In this case the two stations should be placed on opposite sides of the channel, and the text-books (see Stotherd's "Submarine Mines") then recommend that the mines should be laid in rows across the channel, and fired by separate intersections, a separate core for each mine being provided between the two stations, and each electric circuit having two breaks in it, one at A and one at B (see Fig. 74). The plan requires a large amount of multiple cable (as a separate core for each mine is required between A and B), and is consequently but seldom used except for firing a few advanced

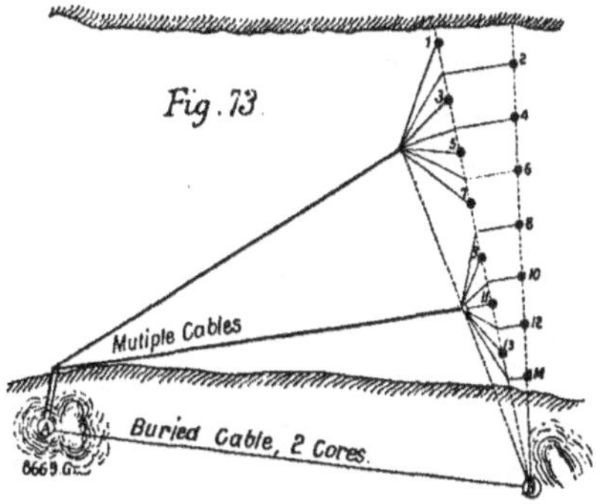

Fig. 73.

and scattered mines, which need not be moored in rows. The instruments employed in our service for firing mines by horizontal cross intersections, are termed "telescopic observing and firing arcs," and are fully described with minute drawings on pages 228 to 231 of Stotherd's

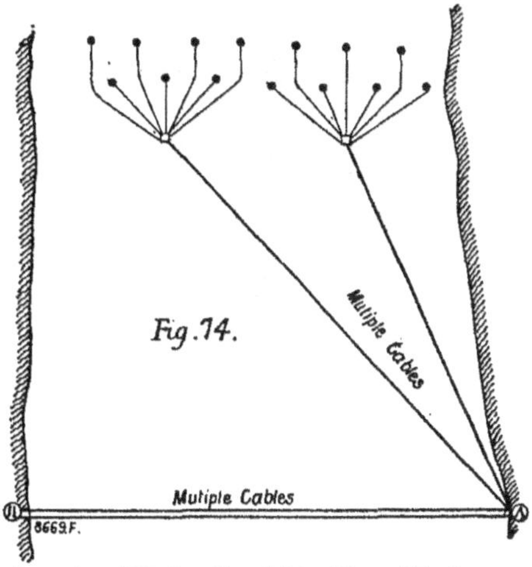

Fig. 74.

"Notes on Submarine Mining," sold in New York. A general idea of the instruments may be formed from the following outline description. The "firing arc" employed at station A consists of a cast-iron skeleton frame forming 77 deg. of a circle of 3 ft. 6 in. radius. There is a levelling screw at each corner. The front arc is so arranged that several insulated metallic foresights can be clamped to it. Each of these is situated in the line of sight to a distant mine, and each is connected by an insulated wire with the cable core leading to that mine. A telescope is mounted with its vertical axis to pivot on the centre of the circle. A light arm is fixed to this axis, and sweeps round with the telescope and just under it. The arm carries a contact-making point at its outer extremity, which is connected to the firing battery by an insulated wire secured to the arm, and leaving it at the inner end. The lead to the firing battery passes through a key under the control of the observer. In case the front sights should slip after the mines are laid, the telescope is provided with a horizontal graduation and vernier. The telescope is set on a distant fixed point and the reading taken, and as each mine is laid the reading is taken. The correct positions of the sights should be tested daily by means of the graduated arc. By these means the fore sights can be fixed afterwards at any time in their correct positions. For short ranges the telescope can be removed and the observing done through the sights. Suitable motion in azimuth is given by a milled head and gearing. The arc for the station at B is much narrower, because one sight only is used. In other respects it is of similar construction, but there is no sweep arm, the electric circuit being closed by a simple finger key.

These instruments are not intended to be semi-automatic in their action, as might be supposed from their construction. They are not sufficiently accurate in their mechanical construction for that. They are used thus: As soon as a vessel arrives upon the line of mines, as seen from station B, Fig. 73, the observer presses his firing key during the whole period that any portion of her hull is on the line, thus "earthing" the firing battery at A.

The observer at A moves his telescope until the line of sight covers the nearest mine to the line of the vessel's advance, and if the vessel come near enough to this line of sight he presses the firing key. Should this occur at the same time that the firing key at B is pressed, the mine is fired. There are several defects connected with this system:

1. The firing connections are exposed to the open air and weather, as the telescope cannot be worked accurately behind a glass screen. Rain, dust, or snow may consequently cause difficulties at critical moments.

2. An electrical leak in the core connecting A and B causes a mine to be fired when the observer at A alone is pressing his firing key.

3. Two instruments and two observers are required for each row of mines, and as each observer should have an assistant to take his place in case of casualty, a large number of trained observers are required.

4. The construction of the firing arc is not suited for observing

through a small aperture, which is now advisable on account of the development of machine-gun fire.

Plane Tables.—Several systems depend upon the plane table for their general application. The earliest forms were very complicated and intricate. The one brought out many years ago by Messrs. Siemens and Halske, of Berlin, and since brought to a high degree of efficiency, consists of a plane table placed in any secure position, and carrying a chart to scale of the mined waters. At the positions of two observing stations A and B on the chart at A, see Fig. 74, are pivotted vertically two light aluminium arms which sweep horizontally over the chart, one slightly over the other, when actuated by electric currents that move a system of electro-magnets in connection with them.

The currents are produced by the motions of magneto-induction apparatus operated at each of the observing stations by the same winch handle that moves the telescope in azimuth. It is so arranged that each telescope and the aluminium arm with which it is connected electrically in the chart-room are always parallel to one another in their projections on the horizontal plane. Consequently, when the telescope points on a certain actual position, its aluminium arm points in the same position as indicated upon the chart. If the two telescopes are pointing on the same object the intersection of the aluminium arms projected upon the chart indicates the position of the object on the chart. If the mine positions are plotted on the chart the observer in the chart-room can, therefore, at once see whether a vessel, which the two telescope observers are following, crosses the position of a mine, and if a firing key for that mine be at his hand he can fire the mine at the correct moment of time. With this system a large number of mines can be operated by two instruments. But this is a positive disadvantage when several vessels attempt to rush through the mined waters nearly simultaneously. Electric caution bells should be rung or other signals made in the chart-room by the distant observers, in order that the chart observer may know when the telescopes are on a vessel; and the greatest care is necessary to prevent the observers directing their telescopes on different vessels, or on different parts of the same vessel.

The instruments are very costly, and some arrangement which is less complicated and difficult to keep in order is preferable. The principal merit of the plan lies in the fact that the firing arrangements are under the control of an operator in a secure bomb-proof, and that the observers have simply to use their telescopes and ring or not ring the caution bells which they control.

The following arrangement, which is much simpler, has just been patented by the writer. The mines are usually arranged in two rows to converge on an observer at B (see Fig. 75). If four rows are required a second advanced observer at C is required. The observer at A is situated over a bomb-proof, and the telescope can be fixed to a vertical axis carried down into the said chamber. This vertical rod or tube carries a light horizontal arm near to its lower extremity, so that the arm and telescope are traversed simultaneously. The arm sweeps over a plane table with a scaled chart of the mined waters upon it. The destructive area of each mine is represented by a small platinised metal button, which is connected by an insulated wire with the mine itself. The arm carries a straight-edge contact-maker, a stretched wire forming a convenient arrangement for this purpose. The straight edge is divided

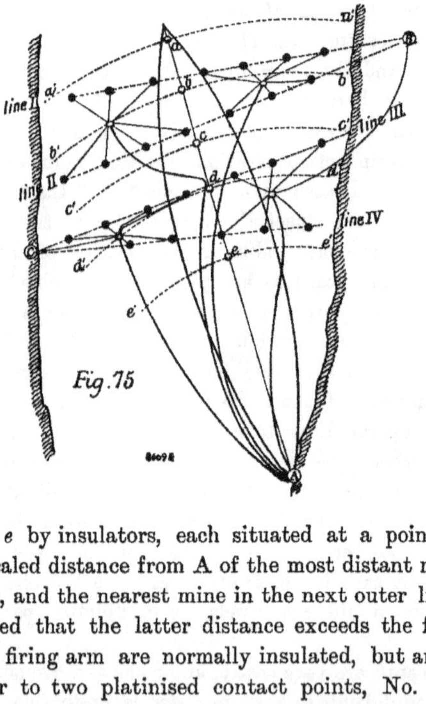

Fig. 75

at *a*, *b*, *c*, *d*, *e* by insulators, each situated at a point intermediate between the scaled distance from A of the most distant mine in each of the inner lines, and the nearest mine in the next outer line; the mines being so planted that the latter distance exceeds the former. These sections of the firing arm are normally insulated, but are connected in pairs, each pair to two platinised contact points, No. 1 and No. 2, forming part of a small instrument illustrated on Fig. 76. A tongue *t* normally midway between these two points is platinised, and is connected to the firing battery F B. The tongue *t* is carried on the extremity of an armature mounted on an insulated axis *o*. The armature consists of a small flat permanent magnet, with a soft iron bar rivetted on either side of it. The regulating springs R R with set screws S S keep the armature in a central position between the horns of an electro-magnet M N when no electric current is passing. A small battery to "earth" E at station B (or C) is connected to a single core leading to A by two double contact-makers, so that a pull on No. 1 cord (Fig. 77) will actuate tongue *t* at A to stud No. 1, and a pull on cord No. 2 will move *t* at A to stud No. 2. The observer at B or C therefore simply pulls cord 1 during the passage of a vessel across line 1, and cord 2 during her passage across line 2.

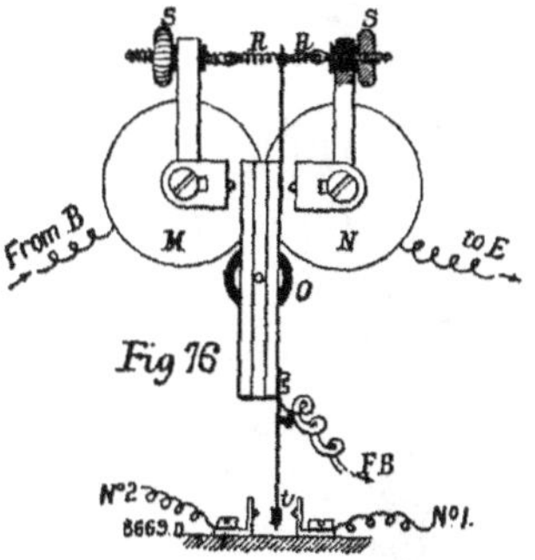

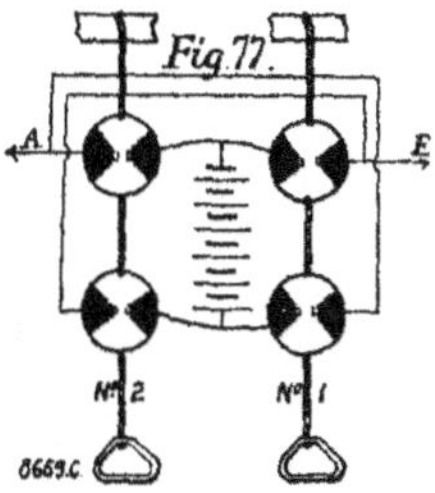

The observer at A keeps his telescope on the vessel, and pulls a contact-maker with his foot or knee so long as the crosshairs of his telescope point fairly on the object. If the vessel come within the sphere of action of a mine it is consequently fired at the right moment.

The above arrangements are capable of several modifications. For instance, two cores may connect A and B, and two electric bells of different tones may be rung from B, one when the vessel is passing line 1, the other when she is passing line 2. In this case there should be a chart-room operator who would prime that section of the straight-edge which fires the mine in the line corresponding to the ringing bell. The observer at A would simply follow the vessel with his telescope, informing the chart-room operator in some manner (electrically or vocally) when he is on or off the object. Again, the electric communication between A and B may be broken, or never made, and visual signals from B would then be resorted to.

Instruments of the kind to be erected in a hurry, should have the plane table and chart placed immediately under the telescope, the whole being set up by three levelling screws on the top of a parapet. Moreover it may then be desired to avoid the use of the auxiliary instrument, see Fig. 76, and perhaps to rely upon visual signals from advanced observers. Under such conditions the mines can be planted so that the lines are directed upon any suitable points I., II.,

III. IV. (see Fig. 78), which can be situated on either side of the channel. Thus a red flag by day or a red light by night displayed by observer I., would indicate a vessel on line I. Ditto at III. for that line. A green flag or light at II., and also at IV. At night a white

Fig 78.

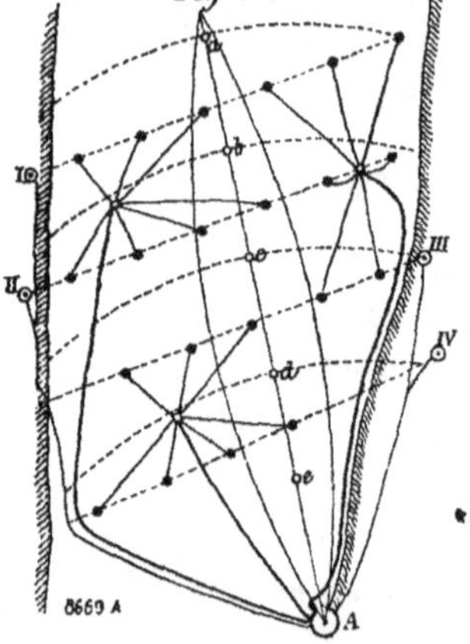

light at each of these advanced posts, and screened towards the front, would be directed continuously on A to indicate their positions to the central observer. If wires be laid to I., II., III., IV., subsequently, it can be done as shown on Fig. 78.

I prefer to use only two lines, as a general rule, in connection with one plane table, because a sufficient number of mines can usually be planted in two lines, for one central observer to attend to, and his actions would be hampered if he or his assistant had to attend to several outlying observing stations. When four lines are used it is therefore better to employ two plane tables at A, each having two lines plotted upon it. When this is done in connection with dispersed observers to the front, the further the two lines on each plane table are separated the better, because there is then less liability to make mistakes. Consequently one plane table should take lines I. and III.; the other, lines II. and IV., and the advanced observers be placed accordingly.

It may sometimes occur that mines have to be planted behind the central station A, in which case the same arrangements are applicable, the outer observers being located towards the rear of the general position.

When a channel is very broad, each side may conveniently be dealt with separately, the mines on each side channel being planted in lines converging in the other station. Thus, in Fig. 79, the mines in the right channel would be laid in lines converging on station B, and within the ring areas described from A as centre. Also the mines in

Fig. 79.

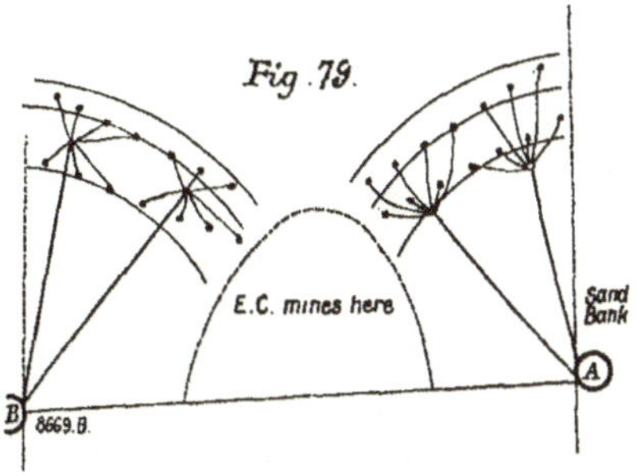

the left channel would be laid so as to converge on A, and would be connected to a plane table in B. The central portion of the channel could conveniently be mined with electro-contact arrangements.

It will be noted that in this system of firing by observation, the angles of intersection do not vary greatly from the best possible angle, viz., 90 deg.

In conclusion, it should be observed that whatever arrangement be used for firing by observation, it should be simple, not liable to get out of order, well protected, effective when the vessel of a foe is enshrouded in smoke. Every arrangement possesses certain inherent defects and advantages. It is for the adept to select the one which he considers best adapted for any required conditions and locality.

CHAPTER XII.

THE FIRING STATION.

THE number of wires unavoidable in every important firing station necessitates a methodical system of connecting and placing same. This is especially the case in our service, where the firing and testing gear is intricate. An excellent system has been elaborated at Chatham under the able directions of Captain G. A. Carr, R.E. A description is not permitted, and if given (with permission) it would simply bewilder the general reader. In fact, it requires a long and careful training to understand the details of an English test-room. Moreover, the details do not apply to the arrangements hereinbefore advocated, but the general idea of method, to avoid confusion, is applicable to all systems, and should be adopted equally for simple as for more complex plans.

Several sets of the apparatus, shown on Fig. 70, page 137, each on its own board, can be fixed on a large deal board secured to one of the walls of the room. On either side of them a batten of kiln-dried teak 9 in. or 10 in. wide, and 3 ft. or 4 ft. long, can be secured, and a number of brass terminals fastened thereto. The cable ends from the mines, or from the observing stations, or from the telegraph stations, in fact all electric wires connected with the firing station, can be brought into the room at one of the corners, preferably near the roof. Each core should then be identified and labelled, and be led to one of the terminals, a small descriptive ticket being gummed on the batten close to the core. The poles of the batteries, &c., should be carried in like manner to terminals on this universal commutator. The various connections can then be made easily and expeditiously. The lead from the firing battery should be kept at a distance from the other leads, and need not be taken to the commutator. For the arrangement described it is advisable to carry the firing lead to the ceiling of the room, and there connect it by means of branch wires to the pull contact-makers, one for each set of fuze signalling apparatus.

The "Earths."—The "earth" used at the firing station should consist of a length of 2-in. steel wire mooring rope carried to low-water mark in a deep trench and immersed in the sea for a length of three fathoms.

For the arrangement described one "earth" can be used for all the batteries, but it is desirable to have a separate "earth" for the resistance test, and the wire rope used for this purpose should be carried to the sea in a trench separated as far as possible from the firing "earth." This, moreover, gives the power to test the joint resistance of these two earths. The inner ends of these ropes should be soldered to a copper strip, the wires being laid out in fan-like form, and carefully soldered to it. An insulated wire or wires can then be led from each "earth" to the battery pole or other point to be "earthed."

Insulated Wires.—The connections in the firing station should be made with a light insulated wire which is flexible, and not likely to break when bent. A good core of the kind is formed of three No. 26 copper wires stranded and insulated with gutta-percha or india-rubber, and with a primed tape wound spirally upon it.

Testing the Firing Battery.—One of the most important tests to be taken at the firing station, is that of the efficiency of the firing battery. This subject was most scientifically investigated by Captain (now Major-General) E. W. Ward, R.E., in the paper already referred to, and which was published in 1855 in vol. iv., R.E. Professional Papers. At that

Fig. 80.

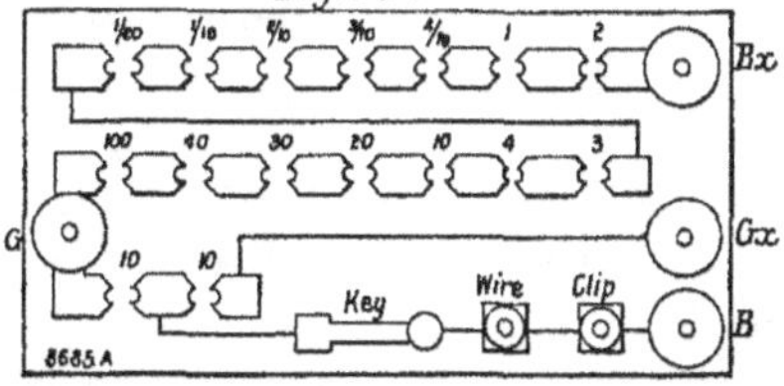

time he invented the instrument since termed a thermo-galvanometer, the fusion of short lengths of fine wire held between metal clips being effected by the battery under examination, and a Wheatstone rheostat inserted in the circuit. The rheostat is a somewhat clumsy and unsatisfactory instrument, and resistances that can be plugged in a box in the (now) ordinary way are preferable. The box should contain a range of about 200 ohms, and a good instrument of the kind suitable for sea mining purposes which Messrs. Elliott are about to manufacture is as follows: The box contains resistance coils ranging between $\frac{1}{20}$ ohm and 211 ohms which can be taken in steps of $\frac{1}{20}$ ohm at any point. A $\frac{1}{4}$-in. clip for holding the wire or wires and a finger key are interposed in the circuit between terminals B B. Two additional resistances of 10 or 12 ohms each are added for balancing by Wheatstone's bridge when required (see Fig. 80), and two terminals G G for the galvanoscope connections. For such a test the unknown resistance is connected to x x.

The resistance coils are made of wire sufficiently thick to give correct results with currents from the firing batteries, prolonged contact at the key being avoided. When testing a firing battery the resistance first unplugged should be less than the estimate, and the wire is fused before the battery has time to polarize. An inspection of the fusion will assist the operator in his next estimate. After a few trials the extreme limit of power of the battery is determined for fusing one wire. Similarly the limit of the power of the battery to fuse two wires in the clip in multiple arc is found.

Then if C denote the current required to fuse one standard wire (the wire employed usually represents that which is employed in the fuzes), and if R and R^1 denote the external resistances found in each test, and if L denote the liquid resistance of the battery under trial, and if E denote the electromotive force of the battery; we have

$$E = C(R + L) \text{ by first test (one wire),}$$

and

$$E = 2C(R^1 + L) \text{ by second test (two wires).}$$

$$\therefore R + L = 2R^1 + 2L \text{ and } L = R - 2R^1.$$

The approximate resistance of the standard wire at fusing point should be determined previously by other processes. This should be added to the unplugged resistance in the first test to give R; and one-half of it should be added to the unplugged resistance in the second test to give R^1.

C should also be a known quantity, and as $E = C(R + L)$ we can calculate the electromotive force of the battery directly its liquid resistance L is found.

But a special instrument for testing resistances by means of currents up to about 1 ampère is not required when the firing station is provided with one of my resistance coil arrangements illustrated on page 137, a clip for the wire bridge connection being added to the instrument at the infinity plug brasses and the resistances up to the 100 ohms plug being formed of wires that do not become heated by short currents up to 1 ampère. By these means the same instrument can be employed for all purposes.

The wire clip can form part of the instrument or be separated, as desired. In the latter case an extra terminal is required connected to the brass of the infinity plug.

Testing the other Batteries.—The other batteries can be tested readily by means of a handy little instrument invented by the late Mr. E. O. Browne, assistant at the Chemical Department, Royal Arsenal, Woolwich. It is a small vertical detector with gravity preponderance for zero and having three coils of 1000, 10, and 2 ohms resistance respectively, any one of which can be used, a plug commutator on the top of the mahogany case being provided for this purpose. A powerful permanent horseshoe magnet is usually employed for controlling the deflections. This magnet should be placed with its axis in line with the axis of the needle at a certain defined distance behind the galvanometer. The dial of the instrument should be graduated thus: A battery of twenty low resistance Daniell cells should be placed in circuit with the 1000 ohms coil, and the deflection taken and marked. The current is then reversed and the deflection taken on the other side of the dial. The battery is then reduced cell by cell, and the deflections marked and numbered to accord with the number of volts which produce them. Finally, readings are taken with the 2 ohms coil, and one cell in circuit. The mean reading is compared with the mean reading produced by the same cell through the 1000 ohms coil, and this comparison is usually called the constant of the galvanometer. It is generally about six in the instruments made by Messrs. Elliott Brothers. Whatever it be, call it M; and if d be the deflection on the 1000 coil and D that on the 2 coil, then $M\,d = D$.

This result being obtained with a cell of low internal resistance L and a potential V, we have

$$V \div (2 + L) \text{ ampères producing } D,$$

and

$$V \div (1000 + L) \text{ ampères producing } d.$$

If a resistance x be now inserted in circuit, we have a current

$$V \div (2 + L + x) \text{ producing a deflection } D^1,$$

and a current

$$V \div (1000 + L + x) \text{ producing a deflection } d^1.$$

If x be small as compared with 1000

$$d = d^1 \text{ and } D = M\,d^1.$$

And as

$$D \text{ or } M\,d^1 : D^1 :: 2 + L + x : 2 + L$$

$$x = (2 + L)\left(\frac{M\,d^1}{D^1} - 1\right)$$

If, therefore, we know the original liquid resistance L of a low-resistance cell producing $M\,d = D$, we can, by means of the above formula, discover any alteration in resistance producing some inequality between $M\,d^1$ and D^1. Moreover, the formula applies equally whether x be an internal or an external resistance to the battery cell, or whether it be added or subtracted, its relative value being *small* as compared with 1000. The formula is not absolutely accurate, as the foregoing indicates, but it is sufficiently accurate for all practical purposes when the resistances to be measured are low, and it is especially useful for finding approximately the internal resistances of small voltaic batteries.

Their relative potentials are also found approximately by comparing d and d^1, the deflections on the 1000 ohms coil. If a fall of one-tenth of the potential occur in a battery, it should be seen to. The various batteries set up for a system of mines should be tested daily, for resistance and potential; and the results recorded in tabular form on a book kept for the purpose.

Testing a Shutter Apparatus.—When a shutter signalling apparatus is employed (see Fig. 69, page 134), it is desirable to test same daily by dropping the shutter (taking care that the firing battery is not plugged) and then plugging a clip and one or two low-resistance cells to the ends of wires W W to note whether the fine wire is reddened. (The small battery and ring bell should be plugged out of circuit.) This tests the efficiency of the spring contacts. The results should be recorded in the book.

The Shutter Adjustment.—The adjustment of the tension of the armature's controlling spring *c* (Fig. 69) depends upon the electrical sensitivity of the shutter apparatus, and this depends not only upon the number of convolutions in the coils and the distance of the poles from the armature, but also and to a great extent upon the mechanical arrangements of the shutter itself. In the apparatus shown on Fig. 69 the lever is arranged so that the longest portion is towards the armature, a high mechanical sensitivity being thereby obtained. The preponderance of the index end should be kept low, the contact friction which impedes the movement of the armature is thereby minimised. The contact points must be kept bright and scrupulously clean, this being seen to daily, and oftener if a stove or a smoky fire is in the room. For this reason it is desirable to keep the room dry by hot water or hot-air pipes which produce no dirt and dust. When each mine contains a testing apparatus, a current from the signalling battery is constantly passing through each shutter coil, and the strength of signal (*i.e.*, the current that causes the shutter to fall) is consequently the difference between the normal current and that produced temporarily by the circuit-closing arrangement in the mine when it is subjected to mechanical shock. In the arrangements proposed in a previous chapter herein this matter is simplified, because the branch cables to the mines are all insulated until a circuit-closer is actuated. The strength of signal is then the whole current produced by the signalling battery passing through the external resistances in circuit, viz., earth at home, cable core, electro-magnet coils at circuit-closer, fuzes, and earth abroad. This current will vary according to the distance of the mines from the firing station, and the consequent variations in the line resistance. When a line becomes leaky, due to faulty insulation, there will be a continuous current through the leak from the signalling battery. In such case the strength of signal is equal to the difference between this current and that produced when the mine circuit-closer acts.

It is necessary to provide a separate signalling battery for a leaky line, separating the group from the remainder of the system served by the same shutter apparatus of seven indices, but the leak should be seen to at once and repaired, and this complication removed as soon as possible. With good gear and good care leaks seldom should occur, and the moment the resistance tests show the commencement of a leak the mine field working parties should attack it, for it is almost certain to develop rapidly, and perhaps make the whole of a group useless. It is generally produced by the chafing of a cable near to or at a cable grip, especially when the mines are subject to the swaying motions caused by strong currents of water.

Having discovered by experiment the minimum mechanical sensitivity at which a shutter should be adjusted, it is easy to discover what current through the coils will just release it, and as the shutter coils are similar to each other, this same adjusting current can be used for all of them.

Each coil should be tested daily in this manner, and the shutter sensitivity altered if necessary by means of the regulating spring *c* (Fig. 69), until it accords with the adjusting current. It is well to commence each test by trying a rather smaller current, gradually increasing same until the shutter falls. The current durations given by a key should be short, or residual magnetism in the soft iron cores of the electro-magnet will cause difficulties.

All this is rather harassing, and the shutter battery is found to give a lot of trouble. The employment of a system in which no shutter apparatus is required, as already explained on page 136, is therefore an important improvement.

Testing the Instruments.—The electrical instruments employed in a firing station should occasionally be tested. The resistance coils should be balanced against one another. The plugging brasses should be frequently cleaned. Each galvanometer or galvanoscope should be tried with a given current to see that its sensitivity remains unimpaired.

Testing for Insulation.—After the mines are laid, it is seldom, if ever, necessary to test a line for insulation, the resistance test, as recorded by a set of resistance coils, being a sufficient test for sea mining efficiency. As has been stated, however, some of the main cables may with advantage be laid permanently *in situ*, their far ends being insulated. These should be tested for insulation periodically—say every quarter. Moreover, the other cables stored in tanks at the store depôt should be tested similarly.

In our service elaborate tests are the fashion, a condenser $\frac{1}{3}$ micro-farad, and a reflecting galvanometer, forming part of the equipment of each important submarine mining depôt.

No useful purpose is served by obtaining the exact insulation resistance per knot of each piece of cable stored in the cable tanks. A much rougher test for insulation is sufficient, and direct testing by Wheatstone's bridge being thoroughly understood by many of the men, it is better to employ it for these tests. This can be done if each depôt is supplied with a graphite resistance of 1 megohm, and two sets of resistance coils, each of 10,000 ohms. A sensitive astatic is then all that is required, and it is very useful for a number of other purposes. This galvanometer should be portable, and easily set up anywhere on a steady and level platform.

Sensitive Astatic Galvanometer.—An excellent instrument was made by Messrs. Elliott Brothers in 1880 from a design and directions given by the author. It has been used for many purposes since, and is an exceedingly useful and trustworthy instrument. It has a fibre suspension of fine silk, and the needle is supported by a mechanical contrivance when the instrument is not in use. It can then receive rough usage without fear of damage. Astatic galvanometers made with an agate and point suspension are constantly getting out of order, the inertia of the needle breaking or cracking the agate plate when the instrument receives a shock.

Not unfrequently these galvanometers, although carefully packed, are damaged in transit and arrive at stations on the other side of the world in a useless condition, and are not easily repaired locally. The fibre suspension galvanometer is free from this defect. It is mounted on a square ebonite block. Its resistance is about 4000 ohms. It has a sensitivity of 12 megohms per volt (that is to say, 1 volt produces 1 unit of deflection through a resistance of 12 megohms), when no controlling magnet is used to bring the needle quickly back to zero, and with the controlling magnet it can easily be adjusted to give 8 megohms per volt. (The maximum sensitivity of the astatic used for submarine mining in our service is only $\frac{3}{4}$ megohm per volt.)

The instrument is provided with a brass cylindrical cover and glass top. This can be removed, if desired, when testing, and small wooden stops placed on either side of the needle so as to prevent undue motion and loss of time thereby. The whole is contained in a small box about 6 in. cube.

The Megohm Resistance.—Many years ago Mr. Johnson, of the well-known firm of Johnson and Phillips, electrical engineers, published a description in the *Philosophical Magazine* of a graphite resistance formed by a pencil mark on a block of vulcanite, and on September 2, 1879, he wrote me a letter on the subject, wherefrom the following extracts are taken:

"My first idea was to make them on vulcanite, because of the comparative ease with which the resistance was adjusted to the required value. After a line had been made with a moderately hard pencil, and when the shellac varnish with which it was covered had become hard, it was easy with the point of a sharp knife to scrape away the plumbago so as to reduce the width of the line until the required resistance was obtained, but my experience has been that those made on vulcanite do not stand well, and I have attributed this to the sulphur working to the surface, as we know it does, and so destroying the continuity. But with glass, although we have a good permanent resistance, the power of adjustment by scraping away the width of the line is lost. The resistance has, therefore, to remain whatever it may turn out to be when first made." . . . . . "One point I have noticed which is very curious . . . . viz., that as the battery power is increased the resistance falls slightly, that is to say, with 500 cells the resistance would be perhaps between 1 and 2 per cent. lower than with 200 cells, but when the ratio of this increase has been carefully measured, it is easily allowed for. I have tested several with 1000 cells, and immediately afterwards with 100, and found the resistance had not altered, although with the higher power it gave a slightly lower value. . . . . . A good deal of care is necessary in making them, and one generally has several failures, but by practice I find that it is possible to get pretty nearly the resistance required. For instance, I recently made a megohm; when finished it came out 1.08 megohm."

The method of taking the insulation test of a cable with the above instruments is depicted in Fig. 81, where C is the cable in the sea or

Fig. 81

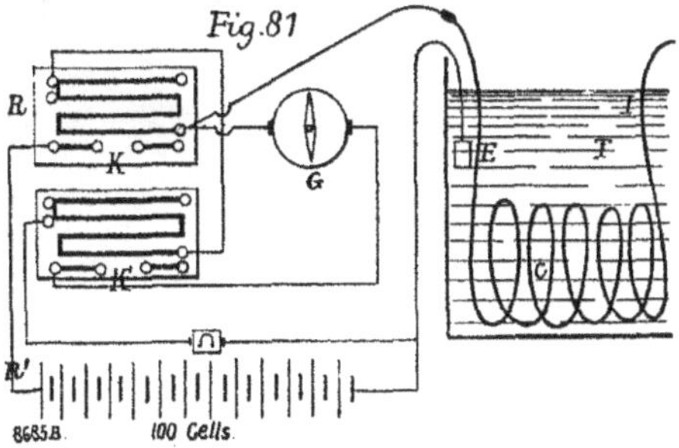

tank T, having the end I insulated and the other end connected through a Post Office pattern set of resistance coils R to the testing battery by the key K. Also, the meohm resistance is similarly connected through

R^1 to K and the testing battery. The galvanometer G is then connected across the bridge by the K^1 as shown, and the test can be taken in the usual way, key K being used for the battery, and key K^1 for the galvanometer.

The resistance R^1 is made 10,000, and R is adjusted until G gives no deflection. Then the insulation resistance x of the cable is found by the proportion

$$R : x :: R^1 : \Omega :: 10{,}000 : 1{,}000{,}000 :: 1 : 100$$

or

$$x = R\,\Omega \div R^1 = 100\,R.$$

Testing the Observing Instruments, &c.—When time permits the electrical parts of the observing instruments, call bells, &c., should be tested for continuity and insulation, all contact points, plug holes, and plugs, &c., being kept bright and clean. Also the leading wires should be examined frequently, and every care taken to keep the whole in thorough working order.

In a paper on "The Electrical Resistance of Conductors at High Temperatures," contributed to the R.E. Institute, 1878, by the author, the following conclusion was arrived at, and the results mentioned by Mr. Johnson (see last page) are probably due to a similar cause, viz., a molecular torsion produced by the electricity. . . . . "The results . . . . appear to substantiate the idea which previous experiments suggested to me, viz., that a conductor heated by a current of electricity offers a smaller resistance than when heated by other and external means to the same temperature." . . . .

"It appears probable that the action may be analogous to that which obtains in an insulating medium when an increase in the electromotive force applied produces a decrease in the resistance of the di-electric."

CHAPTER XIII.

THE STORE DEPÔT.

THE efficiency of many of the arrangements connected with sea mining centres upon good work and good methods at the central store depôt. Numerous considerations, principally connected with the water traffic, prevent mines being laid permanently, so that it is of the utmost importance to arrange so that they can be laid quickly and properly when the order is given to do so. The various stores should therefore be prepared and labelled. In labelling, the best plan is to use numbers, and to keep a record book showing the mine and group to which any number refers.

The Wire Rope should be cut to the proper lengths and each end prepared with an eye and thimble, or whatever the system adopted may be ; a shackle should also be connected to each eye. These mooring lines should then be oiled, number labelled, and put away in batches, those for each group of mines being tied together in one batch.

The Tripping Chains (galvanised) should be prepared in a similar manner, an iron ring added at one end, and a shackle attached. The chains for one group should lie in one heap.

The Junction and Connecting Boxes, the Multiple Connectors and Disconnectors, and all gear of the kind, should be number labelled and arranged systematically.

The Mines, after being carefully tested, should be loaded and stored in a sentry-guarded bomb-proof, with an overhead traveller on the roof and a tramway on the floor. Should it be inconvenient to keep all the mines loaded, a proportion only should be loaded, viz., those to be laid first. The mines should be number labelled.

The Apparatus for each mine should be carefully adjusted, number labelled, and put away in a dry place. The apparatus should not be loaded with the priming charge and detonating fuzes until the order to lay the mines has been given.

The Primary Charges of dry explosive may, therefore, be stored in hermetically sealed metal cases in a store by themselves.

The Fuzes should be stored in a dry place at a distance from any explosives.

The Sinkers may be collected in tiers round a crane close to some portion of the tramway, and not far from the pier.

The Voltaic Battery Cells should be number labelled and stored in boxes ready to be moved to the firing stations at a moment's notice. The salts for same should be stored separately.

The Electrical Instruments should be number labelled, be stored in a warm dry room in glass-fronted cases, and be tested for efficiency periodically, records being kept of same. Some of the less delicate instruments can with advantage be kept ready fixed on the walls or tables of the firing stations away from the depôt.

The Buoys to be used in connection with certain defined groups of mines should be stored suitably and be number labelled. Their mooring lines may be attached to them.

The General Stores, viz., ropes, flags, lamps, &c., can be kept in a large shed suitably fitted and partitioned for the purpose.

The Consumable Stores, viz., the tar, oil, tallow, &c., should be placed in another shed.

The Explosives for the unloaded and spare or reserve mines should be stored at a safe distance from all. An old hulk moored in an unfrequented creek near at hand often affords a convenient store of this nature. Wherever situated, such a store should be carefully guarded at all times, and the explosives be subject to periodical examination, records being kept of same.

The Boat and Steamer Stores, when not on board, should be kept separately from the other stores to avoid confusion.

The Boats should be housed in suitable sheds to protect them from the weather, a boat slip being provided in connection therewith.

The Electric Cables should be cut to the required lengths, their ends crowned and number labelled, and a piece of each core about one yard long left at each end for testing purposes. These ends should be carefully insulated before the cables are placed in the storage tanks. The cable lengths should be stored so that those first required are on the top. Tests for insulation and conductivity should be taken periodically and records kept. The tanks may conveniently be placed near the pier, and if there be a good rise and fall of tide it is advisable to place the tanks just inside the sea wall at such a level that they will fill or empty at high or low tide when a cock is opened in a 4-in. pipe communicating between the bottom of each of the tanks and the sea outside. This saves labour in pumping. The tanks may be made of iron or concrete. I prefer iron, as the concrete tanks are apt to crack and leak, and then give a lot of trouble.

A 15-ft. tank 5 ft. high will hold about 20 knots of single cable, or about 10 knots of multiple, a core of 3 ft. diameter being left for the single, and of 4 ft. diameter for the multiple cable. The dimensions of tanks to contain smaller quantities may be calculated by allowing 40 cubic feet of contents for single cable and 80 cubic feet for multiple cable, in addition to the contents required for the central cores.

The operations connected with the cable laying are laborious. The cables have to be coiled out of the tank or tanks and wound upon drums, which are then transferred to the mooring steamers, and then taken to the mine field and laid.

Messrs. Day, Summers, and Co., of Southampton, now undertake the manufacture of a barge designed and patented conjointly with the author, and in which iron cable tanks surrounded by water-tight compartments, which can be filled or emptied as required for trimming the barge, form part of the structure. Each cable tank and ballast tank can be filled from the sea by a 4-in. cock, and the level of the water in the cable tank can be adjusted as desired by means of a pump, and by altering the buoyancy of the barge by the ballast tanks. A small scale drawing of one of these barges to hold 25 knots of multiple or 50 knots of single cable, is given on Figs. 82, 83. When the cables are stored in

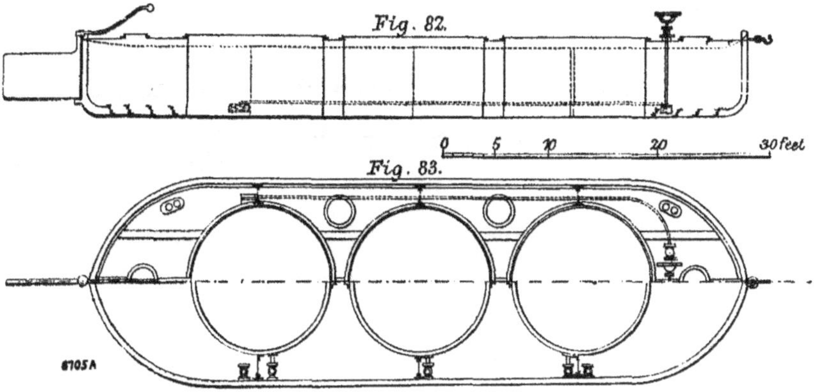

this manner, those which are required for connecting up during the day's work, are removed as usual in the morning or on the previous evening, and the barge can then be towed by any tug, and the larger or longer cables laid out directly from the coils in the barge tanks. In this way many operations are avoided, and the steamers especially fitted for mooring mines can be used for that purpose only. The cable barge is not well adapted for small stations, but is useful and economical in time and labour at large and important stations where a number of mines and cables have to be laid as quickly as possible.

The Pier.—Each depôt must be provided with a wharf or pier fitted with suitable cranes, alongside which the mooring steamers can lie at all times of tide.

A Tramway, 1 ft. 6 in. gauge, with small iron trucks strong enough to carry a load of 5 or 6 tons, should connect the various parts of the store depôt with each other and the pier-head.

It is not necessary to build workshops for artificers except at those stations where experiments or exercise are carried out upon an extended scale and for long periods, but a portable forge, a carpenter's bench, and sets of tools for white and blacksmith, carpenter, fitter, and painter should form part of the equipment at every depôt.

The General Workshop.—One long shed can advantageously be appropriated as a general workshop in which most of the operations requiring cover from the weather can be carried out. Small portions can be partitioned off; one for a storekeeper's office, another for electrical testing, a third for fitting the apparatus, and so on. The size of this, and of all the other sheds, must depend upon the number of mines, the strength of the working parties, &c. A diagram is given on Fig. 84,

Fig. 84.

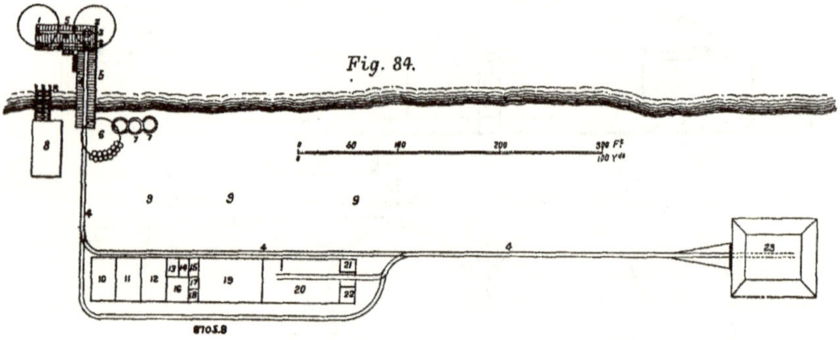

showing the general plan of a depôt for sea mining, but the sites, the stores, and the conditions being so different at various stations it must be treated as suggestive and nothing more.

(1) is a 1½-ton whip hand crane, with a sweep of about 15 ft., placed at one corner of the pier-head.

(2) is a 5-ton hand crane, with same sweep, placed at the other corner of the pier-head.

(3) (3) are small turntables for the trucks on

(4) (4) the 18-in. tramway.

(5) (5) is the pier with steps at the inner angle.

(6) is a 1-ton crane or derrick for lifting the sinkers on the trucks.

(7) (7) are the cable tanks.

(8) (8) the boathouse and slipway.
(9) (9) the parade, available for any open-air work.
(10) general stores.
(11) consumable stores.
(12) boat and steamer stores.
(13) clerk's office.
(14) superintendent's office.
(15) storekeeper's office.
(16) dry store for instruments and other articles.
(17) electrical test room.
(18) electrical fitting room.
(19) general workshop provided with bays and benches.
(20) store for empty cases, buoys, &c., and provided with a bay for hydraulic testing.
(21) (22) shifting and loading rooms.
(23) bomb-proof magazine and store for loaded cases; the tramway runs down an incline into the latter.

The Depôt should be placed in a secure position, and yet not too remote from the mine fields, say not more than three or four knots from the furthest mine, and as much nearer as possible. A small creek running back from the main harbour may often be found, and a site selected so that high ground both hides and protects it.

At an important station the pier-head must be considerably larger than the one shown on Fig. 84, and a third crane should be added so that two steamers can lie at the pier-head and be loaded simultaneously; but it is often preferable to have two moderately sized depôts rather than one large depôt, especially when the mine fields are scattered and separated by considerable distances.

CHAPTER XIV.

Designs for Mine Defence.

The fundamental principles of defence involved in the employment of sea mines must now be considered, and this will lead us to the most interesting work connected with the subject—viz., the designs or chart plans for submarine mining.

They are more intimately connected with fortification than most people suppose. The positions of the forts and batteries, of the mines and cables, of the electric lights, of the firing and observing stations, of the telegraphic and visual signalling forts, &c., should form one harmonious whole.

As no two harbours are alike, so no two arrangements of fortification or of mine defence will be the same; but the same principles apply to all, and they do not differ greatly from the broad ideas that should underlie the preparation of every defensive position. But mining differs from fortification in one important particular. The value of sea mining is greatly enhanced when the positions, or even the approximate positions, of the mines are unknown to a foe. Secrecy is therefore essential. Not concealment as to the efficiency of the apparatus employed, or the manner of its employment; but secrecy as to the waters that are mined. Any artifice which ingenuity can suggest should be undertaken, in order to deceive a foe on this score. Buoys should be laid, which are otherwise useless; bogus mining operations should ostentatiously be conducted for the benefit of spies when time and opportunity are available; false reports concerning the mine fields should be spread; and some of the mines, especially those in advanced positions, may advantageously be laid at night if possible. When drawing up a design, the object of a foe, and the probable manner in which an attempt would be made to attain it, should be carefully considered. It may be the destruction of a dockyard or of a fleet at an anchorage. It may be the reduction of a sea fortress, or the capture of a commercial city or of a coaling station. The attack may consist in a bombardment, or in forcing a channel at speed, or in ascending a river deliberately by "force majeure," including perhaps land forces, as in the American War of Secession—and the defence in each case must be planned accordingly.

Let us commence with the dockyards. Mr. J. Fergusson very truly said in a pamphlet published a great many years ago, and entitled "French Fleets and English Forts," "Turn and twist the question as we may, there is no denying the fact that the proximate and ultimate defence of England must mainly depend on the fleet," . . . the power of which "is wholly and absolutely based on the possession of our arsenals."

A fortified dockyard is not likely to be attacked by land by regular siege made for capture, unless as part of an invasion on a large scale; but a bombardment from land batteries, or from vessels up to perhaps 12,000 yards range, should be guarded against. However, neither ships nor land batteries can effectively bombard an unseen object even at much shorter ranges than 12,000 yards. Consequently, when hills screen a dockyard from view, no position beyond them need be occupied. On the other hand, when ground exists within bombarding range and in sight of the dockyard, measures should be taken to deny it to a foe.

It may be asked,—What has this to do with submarine mining? Let us take an example, and see. Assume that a dockyard is dominated by an island within bombarding distance, and that a bay with a good beach forming a fair landing exists on the outer side of the island.

Evidently, the bay should not only be fortified but mined. A few groups scattered about irregularly between the headlands would be sufficient, inasmuch as their presence would greatly delay a landing and impede the work afterwards until they were destroyed or removed. General Lefroy once said that "When a fleet bombards, the opinion of naval authorities seems to be that the attacking vessels should anchor; if not, and they continue in motion, their distance from the object is constantly varying, much of their fire is thrown away, and they incur numerous nautical dangers," which now certainly include sea mines. But the operation of bombardment is a long one, lasting for many hours, during which the ships engaged in it would, if anchored, be more exposed both to the artillery and the torpedoes (locomotive) of the defence, than if the ships were kept in motion.

Guns cannot be considered efficient against ironclads except at ranges of 3000 yards and under. The forts should therefore be placed some 8000 or 9000 yards in front of the object to be protected from bombardment. If this can be done bombardment is practically prevented, for the most powerful vessels would be much injured by modern rifled guns at battering ranges, and such injuries would necessitate protracted repairs after any conflict with forts, and it is highly improbable that a naval power would risk this loss of efficiency in a fleet at a critical time even for a short period, in order to bombard a dockyard at long range. Sometimes, however, it is impossible to place forts so far to the front, and in such a case a zone of water will exist from which a fleet may bombard and yet be outside the effective battering range of the guns mounted in the forts. Under such conditions groups of large sea mines to be fired by observation should be scattered irregularly in this zone. Mines fired by contact arrangements would probably soon become useless in such exposed positions.

Cherbourg may be taken as an example of an important town and dockyard much exposed to an attack by bombardment. The naval yard, docks, and basins, cover an area 1400 by 900 yards, and the breakwater on which the most advanced forts are situated is only 1900 yards to the front.

Assuming that the artillery mounted in these forts possess a battering range of 3000 yards, it is evident that the vessels of a foe can lie outside this range and destroy the dockyard by deliberate bombardment. A fleet possessing the power to execute such an undertaking is not to be baulked by torpedo boats and small fry of that kind, but a number of large mines placed in these advanced waters would be of priceless value to the defenders.

The attacking forces would then be compelled either to undertake the protracted operations required for the destruction of the mines, or to run a great risk of destruction themselves. In the former case sufficient time might be gained by the defence to summon a relieving force by sea; and in the latter, the results might be felt throughout the war.

Fig. 85 on the next page gives a sketch plan of the arrangements which could be made. The mines are arranged in seven groups, each containing seven mines, and each mine fired singly by observers situated in two stations A and B; the latter, upon which the mines in each group are directed, is situated on a hill 230 ft. high, near the town of Heneville, on the left of the position. The former, on the extreme right of the position, is on high ground, near the quarries of Becquet. The stations are separated by a direct distance of 9000 yards, and the most distant of these outer or advanced mines are about 13,000 yards from station A, and the nearest about 8000 yards off. All these mines can be brought closer in if so desired. On days when the dockyard could be seen at bombarding range, the masts of the vessels could also be seen from the stations through the telescopes of the observing instruments, and the most distant mines could be fired accurately if the instruments are well made, fixed, and served. The observers being on

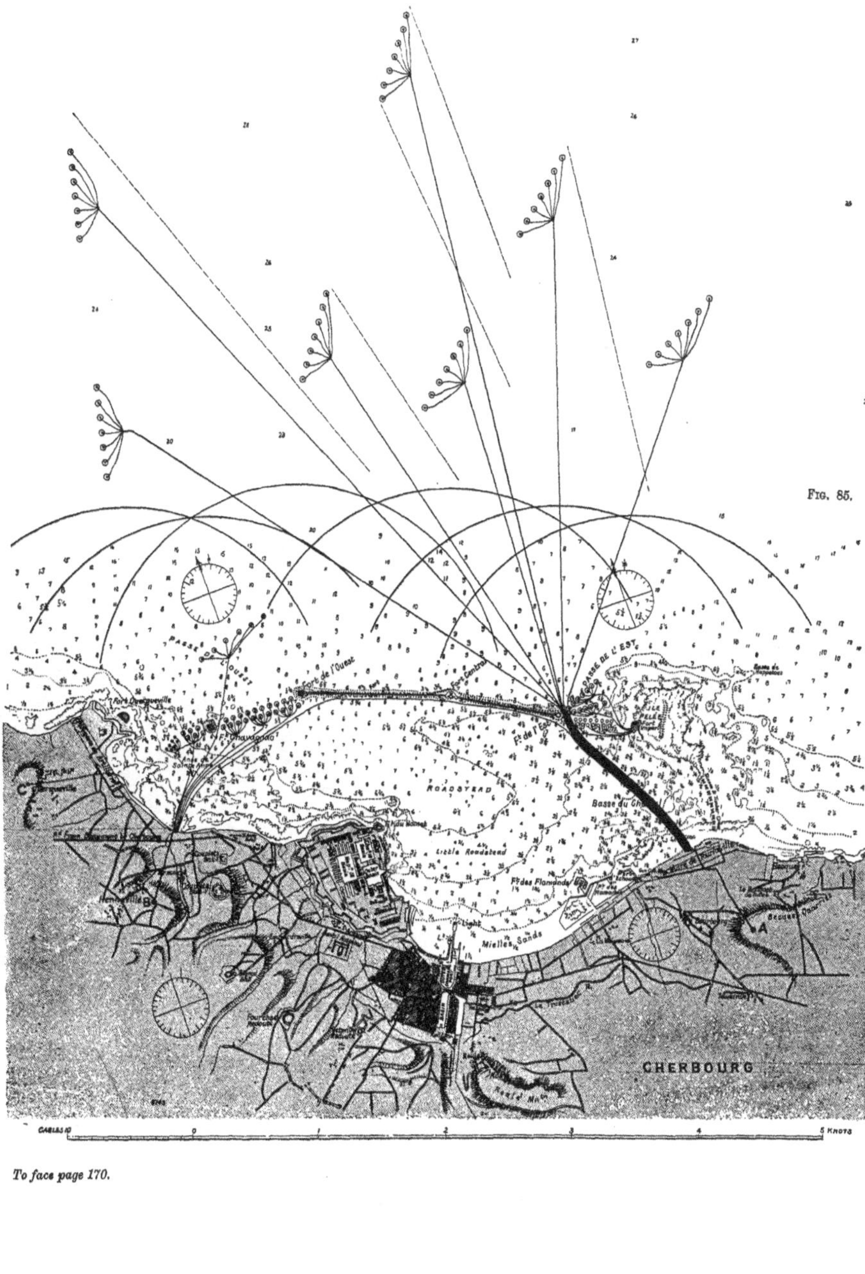

To face page 170.

high ground, are enabled to see at once where a vessel floats, approximately, and the plane table gives them the necessary information concerning the mines. Moreover, the groups are so arranged that no two mines and no two groups are intersected by the same line of sight from station A. The defect which the opponents to double observation firing make so much fuss about is thereby obviated. The arrangement is, I believe, novel, but is only possible when there is a large area to be mined by a small number of groups.

A seven-cored electric cable connects the two firing stations. It is led on plan across the harbour and behind the breakwater. The sea "earth" common to the system would be taken from B, and one core for each alignment would be normally "earthed" at B through a high resistance, and a sensitive galvanometer. The other cores of the multiple cable would be spare, and one would be required for telephonic communication between A and B.

At A the sweep arm of the telescope would be so constructed as to make contact with a series of small metal arcs fixed on the plane table, each covering the angle subtended at A by one group of mines. These metal arcs would be in connection with the cores for alignments, carried to B, and a signal current would pass through the high resistance to earth at B whenever the sweep arm at A touched one of these small metal arcs. In this manner the observer at B would immediately know what group or groups of mines the instrument at A was directed upon, and this although the alignment is itself directed upon B.

The observer at B would deal with the same alignment, or group line, until the galvanometer ceased to deflect, and in this manner it would be hardly possible for the two observers to be looking at different vessels. The other actions would be similar to the operations of double observation firing already described on pages 150 to 153, except that firing by means of metal buttons on the plane table would not be accurate enough for such long ranges, and it would be necessary to construct the instrument at A like a large theodolite with a carefully graduated horizontal circle and vernier adjustment. The telescope could then be clamped with its arm on each mine button in succession, and its line of collimation directed exactly on the mine. On a vessel covering the latter the mine would be fired as soon as the foremast came upon the cross-hairs of the telescope. In order to assist the observer at A in this operation, the mines should not be moored too close together. On the diagram they are shown at intervals of one and a half cables, or 300 yards.

The depth of water does not exceed 26 fathoms at the most distant portions of this advanced mine field, and most of the mines are situated in from 15 to 20 fathoms. Those in less than 15 fathoms can be ground mines containing 900 lb. of blasting gelatine, and the remainder can be 500 lb. buoyant mines moored on a span so as to be submerged about 7 fathoms below low-water level. The cases to be employed and other particulars are given on pages 80 to 88. The diameter of the effective circle of each mine would thus be at least 60 ft.

It does not often occur that advanced mines play so important a part in the defence of a sea fortress, because nations seldom place large arsenals in such exposed positions. When they do so, and spend enormous sums on fortifications and armaments which are powerless to prevent bombardment, it becomes necessary to spend money freely on advanced mines. The system described would cost say 20,000*l.*, which would be reduced by nearly 14,000*l.* if the same number of mines on the contact system were employed in place of the large observation mines, the saving being principally effected by the use of single cable instead of multiple cable, and these mines would be dangerous to a foe at night, whereas the observation mines could not be worked by night at such long range.

However, as before stated, the employment of contact mines cannot be recommended in such exposed positions, and where the tidal currents are strong and the waters turbulent. Such mines would probably not remain long in good order, and repairs might soon become impossible in the advanced mine field if the defenders were weaker than their foe on the open sea.

Semi-Advanced Mines.—A few mines in advance of the forts, and within their effective battering range, should seldom be omitted in the defence of any important sea fortress. The knowledge or the suspicion of their presence in such situation impedes the action of an attacking force immensely; and if any attempt be made to clear a channel by countermining, it has to be commenced from afar, and conducted for a considerable distance, with an enormous expenditure of explosive material, and by operations so tedious and dangerous as to invite disaster. But we must not anticipate. Countermining is now considered so important a means of attack (whether correctly or not) that it deserves something more than a passing remark, and shall be dealt with in another page. If countermining be not resorted to, or the mines rendered inoperative by other means, it is evident that a fleet attacking at battering range must come to an anchor, or verily the vessels would "incur numerous nautical dangers" (Lefroy).

A group of seven large mines fired by observation is shown in the Passe de l'Ouest. They are moored in from 8 to 10 fathoms, and may, therefore, be ground mines containing 600 lb. of blasting gelatine, an effective circle on the surface of rather over 30 ft. radius being thereby obtained. The mines in this group could not act by contact for reasons already stated with reference to the advanced mines, and in addition because they would impede the traffic of the French vessels.

The mines are therefore spaced at about one and a half cable intervals, and are designed to be fired by observation from station B, a cable core being taken to an auxiliary station C on the semaphore hill at Querqueville, and visual signals may be used if the electric communication should fail between B and C. The firing arrangements may be one of those described on pages 150 to 153, say one of the lines on Fig. 78.

More than one group of semi-advanced mines would probably be employed, but the group shown as an example is sufficient to indicate the practice to be pursued.

Mine Blocks.—The anchorage and dockyard of a great naval arsenal like Cherbourg must also be secured as far as possible against direct capture by a fleet. Powerful guns mounted in armoured forts and shore batteries form the chief defence, but they can be passed by first-class ironclads unless obstructions of some kind are added to delay the passage of vessels up important channels. The best obstructions are submarine mines, and when well protected both by heavy artillery and by quick-firing guns, the defence to prevent passage then becomes so powerful, and the operations necessary to force passage become so difficult, tedious, and dangerous that we may feel sure they will not often be attempted.

When mines are employed in this way, they form what is termed a mine block, and it is usual to leave certain portions free of contact arrangements, so that friendly vessels may pass and repass without injuring the mines. These channels should, however, be closed to the vessels of a foe by mines at a lower level, either ground or buoyant (according to the depth of water), and fired by observation. A channel thus mined is absolutely blocked so long as the system remains in good order, but a mine block is somewhat vulnerable to attack by countermining, because the position of the mine block may generally be guessed with an approach to absolute certainty. For instance, at Cherbourg, if used at all, the blocks must exist in the waters between Ile Pelée and the Fort de l'Est at one end of the breakwater, and in the water between the Fort de l'Ouest and the western mainland on the other side. The fact that such mine blocks can be quickly pierced by countermining produces a want of confidence in their efficiency, and inasmuch as they are frequently very costly, owing to the large number of mines required, the advisability of using them in a situation such as the western entrance to Cherbourg is open to question. It may be far better to use the same mines scattered irregularly in waters further to the front, the semi-advanced mines being reinforced and the absolute block sacrificed.

However, mine blocks are the fashion, and we must therefore describe them.

By referring to the diagrams of Cherbourg, it will be seen that the design for a mine block on the eastern side consists of two rows of ground mines fired by observation from Fort Impérial, whence the electric cables are led. The front row of seven mines in 5 to 6 fathoms may be formed with 400 lb. charges spaced at intervals of about half a cable. The row is aligned between the outer buoy and the Fort de l'Est, whence an observer signals to the Fort Impérial. The inner and similar row of mines is aligned between the buoy off Trinity Point and the Fort de l'Est. Behind these lines are placed three or more groups, each of five electro-contact mines in two rows, directed upon Fort Impérial for facility in laying. They are spaced at about 200 ft. intervals, and can carry charges of 70 lb. to 100 lb. as considered expedient. The electric cables from these mines are also led into Fort Impérial.

It will be noticed that these mines are so situated as to be protected, as far as possible, by the breakwater and by the Ile Pelée. They are, in fact, sheltered from all except north-westerly gales. It is assumed that powerful electric search lights are mounted on these forts.

As the electric cables from the advanced mines converge on the Fort de l'Est, and are carried thence to the shore near the Greves battery, it is necessary to prevent boats from attacking same by creeping operations under the cover of darkness. For this purpose a passive obstruction consisting of heavy cribs of timber filled with stones should be placed so as to connect the Pont de l'Ile and the mainland at "la vielle beacon." Greves battery should also be strengthened to resist capture by surprise, and some quick-firing guns be mounted therein on disappearing carriages. This is most important. At present quick-firing guns are mounted on land as they are on ship, upon fixed stands, and consequently both guns and stands would be easily destroyed by the preliminary artillery fire that should prepare the way for any night expedition of the kind. If the quick-firing guns were mounted on disappearing carriages, neither guns nor carriages need be exposed or their presence known until the time arrived for using them. I pointed this out immediately after the operations at Langston Harbour last year; as also the absolute necessity to employ smokeless powder for these quick-firing guns.

Observing station A should be protected by a small fieldwork with wire entanglements and other obstacles, and the same applies to stations B and C on the west. Here the mine block may consist of two rows of ground mines in from 4 to 8 fathoms, and containing charges sufficient to produce effective surface circles of 30 ft. radius. The mines may be spaced at half-cable intervals, and be fired from station B, whence the electric cables would be led. The rows converge on station C where an observer would signal to B when a vessel crossed one or other of the rows of mines. Or the signals may be made from Fort Chavagnac which exists upon the alignments. A cable core is taken to the fort for this purpose from the multiple main cable leading to the semi-advanced group. The shallower waters between the 3 and 5 fathom lines are closed by a group of seven electro-contact mines, the firing being under control from station B. It will be noticed that a small space of unmined water exists between the observation and the electro-contact mine. This is unavoidable, because the former destroy the latter if placed too close to them. The electric cables from this mine block are led across waters which are not required for anchorage, and they are protected both by the forts in front and by the mines themselves.

The junction box for a group of mines should be situated as near to the mines as possible, both to economise electric cable, and to facilitate repairs. When two or more rows of mines are close together, the junction boxes can be placed as shown on the design for the left mine block, but when the mines are arranged in single rows the arrangement shown on diagram for the advanced and semi-advanced mines must be followed. The general sea defence of a fortress like Cherbourg resolves itself into:

1. The protection of the dockyard and town from capture or bombardment.

2. The protection of the anchorage.

Not only is Cherbourg extremely vulnerable to bombardment, but it possesses another weak point, enabling an active foe to capture it by assault without bringing a single vessel inside the breakwater. The precise method need not be further alluded to, except to note that the same mines which throw difficulties in the way of bombardment, would also increase the dangers run by vessels assisting in the operation of an attempt at capture by assault. The anchorage may be considered secure; the powerful forts in front of it sweep with their guns, at what may now be considered short ranges, the waters within engaging distance of the roadstead, and make it highly improbable that any foe would send a fleet, or any portion thereof, into a position of such nautical danger until the defence had been broken down by the capture or destruction of some of the forts and batteries. As regards boat attacks, or what will in future take the place of the old cutting-out expeditions, sea mines are not likely to hinder them much. Two or three boats might come to grief, but the remainder would get through, and be able (if not met by other boats) to attack vessels in the anchorage. The only trustworthy defence of vessels at anchor against boat attack is boat defence, aided, as it would be by search lights and quick-firing guns both from ships and forts.

Reverting to the sea-mining design for Cherbourg, it will be noticed that it is proposed to employ large mines fired by observation rather than electro-contact mines. Reasons have already been given for this choice in the open waters; but contact mines are the general favourites, especially for the principal mine blocks. The preference usually shown for contact mines is not easily accounted for. On the contrary, these mines are so local in their action, that modern men-of-war with their numerous water-tight compartments are more likely to succumb to the racking blow of a large mine. Moreover, the principal mine fields are usually placed precisely where it is best to provide for locomotive torpedoes of the Brennan or similar types, and such weapons can be used over waters sown with observation mines, but not over waters sown with contact mines.

Leaving Cherbourg, where advanced mines are so valuable, we will now turn to an example in which the mines should be placed in retired positions. The great city and harbour of New York may be taken as a case in point.

CHAPTER XV.

Designs for Mine Defence.

Dockyards where men-of-war are built should be secure from attack, and should therefore be situated many miles up a tortuous river or inlet difficult to navigate at the best of times. Chatham is a practical example of this ideal.

The protection of large commercial shipping centres from purely naval attacks is easily effected if they be similarly situated. The great cities of London and New York are typical cases. It is not desirable that our own defences should be discussed in detail, especially by one who for many years was engaged upon them at our War Office. New York will, therefore, be selected, and a design for the sea mine defence will be drawn up and described.

The position of the mine fields should be retired from the open sea both because they would then be more difficult to approach, and because the shelter would enable the miner to carry out the mooring operations in rough weather. Care must, however, be taken to moor some of the mines at or a little beyond the limits of bombarding range, and the remainder should be scattered in groups or fields as irregularly as may be compatible with their protection by light artillery, and especially quick-firing guns mounted in proper emplacements.

Absolute mine blocks which are so fashionable, with their floating impediments telling a foe where the mines are laid, should be avoided. This method places too many eggs in one basket, and shows the position of the basket. A mine defence should be deep and narrow in plan, rather than wide and shallow; and the centre of each channel should be mined more than the sides.

Mines should extend right through the defence, to the very last entrenchment. Sir Lintorn Simmons once said, " A gun for the defence which can be reserved until the attack is in the last period is worth anything" (R.E. Papers, vol. xviii., 1870); and the same remark applies to mines. In a letter to the *Times*, 1855, signed " B.," and attributed to the late Sir John Burgoyne, we read, "one of the principal ingredients in defensive works is an obstacle to the approach of the assailants."

On this principle mines should be moored so as to help the forts when they are attacked, by obstructing those portions of the channel outside the forts at engaging distances.

Small side channels that are not required by the defenders may be blocked by passive obstruction or mechanical mines, especially if these side channels are likely to prove of use to a foe in delivering boat attacks. Let us apply these principles to our example, New York.

This magnificent emporium of trade, whence radiate the pulsating arteries essential to the life of one of the greatest civilised nations the world has ever seen, is situated on a peninsula on the left bank of the River Hudson, and is covered from the open sea by the end of Long Island, between which and Staten Island the Hudson flows through the Narrows into an estuary about seven miles square (49 square miles) containing several channels divided irregularly by large banks with about two fathoms over them at low tide.

Long Island, about 100 miles by 17, covers a large sound or arm of the sea, over 20 miles wide, that separates the island from the States of Connecticut and New York. At a distance of 12 miles from the city this sound narrows down to a width of 1½ miles, and at 8 miles from the city it is only ½ mile wide. Here two forts, one at Willet's Point on Long Island, the other at the extremity of Throg's Neck, on the opposite shore, protect the channel, which from this point inwards is called the East River. Its width is still further reduced as it approaches the city, until finally at Hell Gate it is less than a ¼ mile wide.

Long Island thus protects the city from the sea, obliging any naval attack to be delivered on one or other of two intricate paths of approach. But the defence of an island against a foe who possesses the command of the sea is much more difficult than is that of the main land of a country held by a courageous people and intersected by numerous lines of communication. Long Island, therefore, is at one and the same time a source of protection and a source of weakness. It gives strength to resist a purely naval attack, but is very vulnerable to a combined naval and military operation.

In these days it is impossible to prevent troops landing when the operation is covered by a fleet, and it is also impossible to check their advance so long as they advance in a parallel line and within the effective range of its artillery.

New York is open to capture by an operation of this nature, a strong force landing on Long Island from the Sound as near to the city as possible, and advancing by the East River shore, under the covering protection of a fleet, and assisting the ships by capturing the batteries or mining stations on that shore, which would thus be taken in reverse

With Long Island in its present defenceless state such an operation would be quickly done, in spite of submarine mines, dynamite guns, plunging Davids, and what not. To describe a possible *coup de main* is to suggest a defence, and our cousins might do worse than spend some of their annual surplus in the construction of a string of forts between Jamaica Bay and East River. Taking things as they exist, the following arrangement of mines would give a strong defence to East River against a purely naval attack. In order to hamper any attack on Fort Schuyler the navigable water to the west of Hewlett Point and Elm Point should be mined. For reasons already stated, the firing stations should not be located on Long Island, but on the opposite side. Ground mines can be used in front of the forts, and be charged with 600 lb. or 900 lb. of blasting gelatine according to the depth of water (see pages 80 and 83.) The principal firing station may be situated to the north-west of Fort Schuyler, near enough to be under the protection of the fort, and far enough to be clear of its smoke and of the fire which it draws upon it.

An auxiliary observing station may be placed at M (see Fig. 86), or further from the shore if M be considered too exposed to attack by a party landing from boats at high tide. But its position ought to be screened, and should not be known to a foe, and this remark applies to every observing station used in connection with sea mines.

The mines are shown as moored in four lines converging on M, and the cables would be carried to the back of Throg's Neck to the observing station at that place; this would connect with M by means of a three-cored cable, two cores being required for observing and one for telephoning. Lines 1 and 3 would be operated from one of the writer's plane table observing arcs, lines 2 and 4 from another, and there would be an observer to each core at M, who would send a positive current for one alignment, and a negative for the other, to the instrument shown on Fig. 76, page 151.

These mines being spread over a large expanse of water would be most useful against vessels that might engage the forts at battering ranges.

The water to N. of Willet's Point could also be mined similarly, but Throg's Neck is a position excellently well adapted for a battery of dynamite guns firing both to front and rear, in which event the waters within a radius of a mile can be kept clear of mines.

In nearly every defence some of the side channels are a source of weakness. They should be blocked by mechanical mines, or by passive obstructions, or both combined. Thus two groups of mechanical mines may be placed between City Island and Rodman's Neck, if other considerations do not prohibit same, and another group may be placed off the rocks at Elm Point.

Abreast of Fort Schuyler a mine field may be formed consisting of four groups of electro-contact mines flanking a fairway, mined with several pairs of observation ground mines. This fairway can be in-

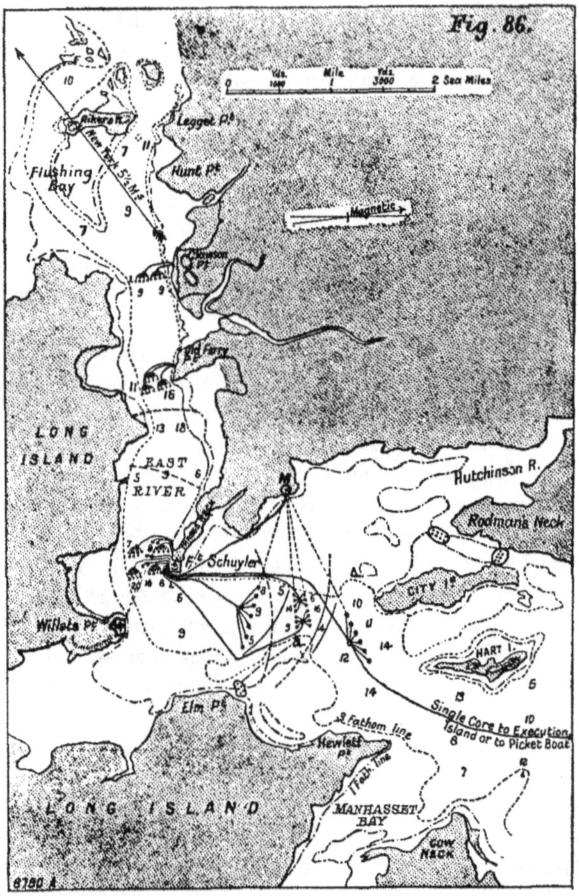

Fig. 86.

clined so that its line of direction falls on Willet's Point, and each pair of mines would be fired from the central station at Throg's Neck when a vessel came on the observed intersection. One of the writer's plane table arcs might be employed for this work.

These portions of the defence may collapse after due resistance, and other mine fields in rear should therefore be provided. One can be placed at Old Ferry Point, another at Clawson Point, and still another perhaps at the Brothers, each having a narrow fairway free of contact mines.

An attacking squadron that succeeded in forcing its way to the Brothers would be within shelling distance of the city, and terms would probably be arranged to prevent further operations.

Let us now turn to the principal commercial entrance to New York Harbour.

The attacking forces would here meet with numerous nautical difficulties. The deepest water over the bar is but 3¾ fathoms at low water, and the rise of tide at springs is less than one fathom—total, 22 ft. 6 in. at low and 28 ft. 6 in. at high water. First-class ironclads should therefore keep outside, and any attack on this side must be made by war vessels of smaller draught. The channels inside the bar are intricate, and skilled local pilots are required to take steamers into port. If some of the buoys and light vessels were only slightly shifted the navigation of vessels would be made so difficult to strangers as to be well-nigh prohibitive. Moreover, the land is so distant and so hard to approach, owing to the flats that extend for miles in front of it, that a simultaneous attack by land could receive no assistance from the forces afloat. Combined operations like those suggested for the advance up the East River are therefore impossible.

Considering these things, it certainly appears that New York, like some other places, has a weak back entrance and a strong front door. Yet an attack *viâ* Sandy Hook and the Narrows seems to be feared more than one in the other direction, if one may judge from the fortifications now existing, especially at the Narrows.

The writer believes that the key of the lock for securing the main entrance to New York Harbour will be found at the inner end of the sandbank called the Dry Romer. This is ten miles from the nearest point of the city and eight from Brooklyn. The Narrows are only six miles from the city and four from Brooklyn, and vessels lying outside would be within bombarding distance of them both. Every effort should therefore be made to present an effective resistance to an attacking squadron before it comes so far.

The Swash Channel joins the main channel close to the north end of the Dry Romer, the navigable water being only 1250 yards wide at this point. It is bounded on the west by the Staten Island flats, with an average depth of only two fathoms over them.

The East Channel is also 1250 yards wide at the north end of the Dry Romer. This fine channel, although not much used by commerce, has 3¼ fathoms over the bar, and might be used by the attack in war, as it lies beyond the effective range of Sandy Hook Fort. All the rest of the harbour entrance is forbidden to vessels drawing more than 14 ft. or 15 ft., that is to say, to war vessels that would cross the Atlantic. An ironclad fort on the north end of the Dry Romer would consequently hold this entrance to New York, and with additional certainty if mines were placed in the channels on either side of it. As no fort exists there, the mines are all the more necessary, and some makeshift arrangement should be devised both for protecting them against boat attack and for providing a firing station as close to them as possible. This could be done by floating a strong iron hulk to the spot, and then filling her with sand, leaving chambers on the north side for firing and observing stations, and mounting quick-firing guns on carriages disappearing through the deck, the guns remaining up when in action, and out of sight and protected as far as possible when not in action.

The mines can be arranged in various manners, and the plan shown on Fig. 87 provides for the main channel a combination of electro-contact and of observation mines, the latter being charged with 600 lb. of explosive and moored on the ground in two lines converging upon Norton Point. They are not placed directly across the channel, but diagonally, and so that the cross intersection firing may be effected from the temporary station near at hand. In this manner the west part of the channel becomes a fairway free from contact mines, and available for the traffic of the port. A plane table observing arc can be used, and a single core would be led to the alignment observing station on Norton Point, a second core would be required for telephonic communication, and a third core would be held in reserve, as spare. A single core should be carried on to Sandy Hook Fort, as shown, for communication, and perhaps it may be led into the Swash light vessel on the way. The East Channel can be closed by four groups of electro-contact mines.

In rear of the Dry Romer defence a second series of mines may be moored in the main channel off Norton Point, in order to hamper the attack on the Narrows. These mines should be scattered over a wide area, and observation mines may advantageously be resorted to, because vessels would not attack the Narrows at night. The mines can be charged with 900 lb. of explosive (the depth being about 11 fathoms), and can be moored on the ground in two lines crossing one another, and directed, the one on a station near Fort Tomkins, the other on a station near Fort Hamilton.

Still nearer to, and in front of the Narrows, a further system of observation mines moored on the bottom in two rows forming a re-entering angle can be directed on the two stations last mentioned, and be fired therefrom, by double observation, one plane table observing arc being used at each station and two cores connecting them for firing purposes. A third core of the seven-cored cable shown on the figure can be employed for telephonic communication. Two more cores would be required for the mines off Norton Point, and two cores would be held in reserve as spare. A group of electro-contact mines can be

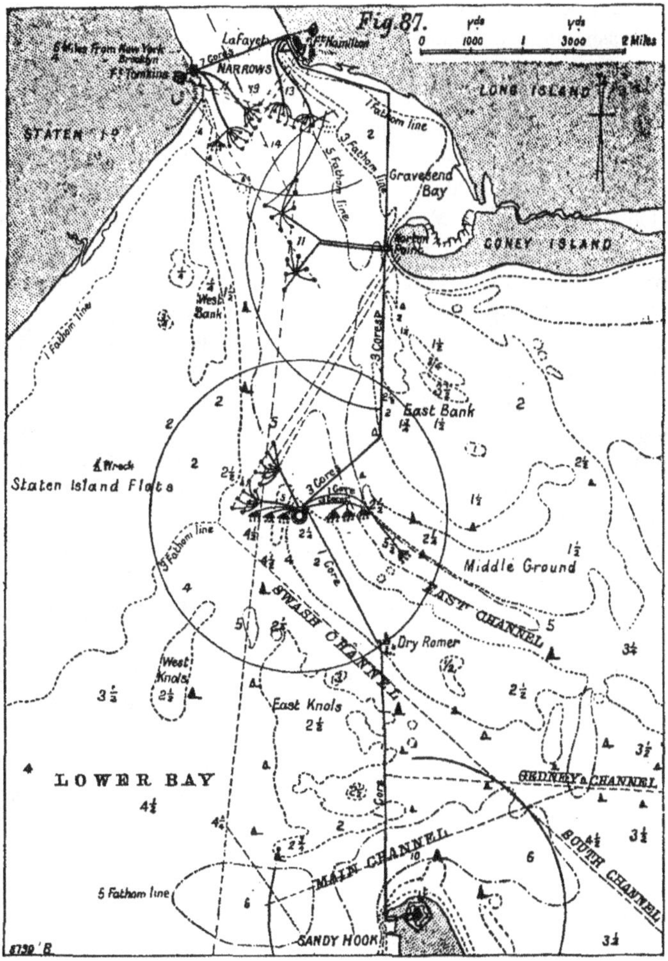

placed on each flank. The water is somewhat deep at the Narrows, and the defence can here be left to artillery and torpedo guns mounted on the heights on either side, and to locomobile torpedoes actuated from suitable positions on either shore.

The defence of New York Harbour offers a very interesting example of the general ideas which govern the application of submarine mines. In every large harbour, however, the possible permutations and combinations are numerous, and no two designs drawn up independently, even by officers who have been trained in the same schools, are likely to be precisely similar.

Thus, in the example before us, many engineers might prefer to sow the Swash and East Channels with mechanical mines, and to place a complete system of electrical mines in the main channel off Sandy Hook Fort, friend and foe alike being thus compelled to use this channel in time of war. Such an arrangement would be strong, and would deny the lower bay to a foe; but the defence would be somewhat disconnected, and for this reason would, I think, be weaker than the one proposed and illustrated on this paper. Moreover, inasmuch as a war may last through the winter months, and masses of ice come down when the Hudson River breaks up, mechanical mines in situations like the Swash would certainly be destroyed by self-ignition at such a time.

Coaling Stations.—The remarks already made on the mine defence plans for naval arsenals and for commercial harbours or rivers, apply also to coaling stations, except that the positions of the latter can be, and generally are, so chosen that their defence requires a much smaller expenditure on guns, mines, and garrisons. A coaling station so situated that its defence would entail a heavy expenditure, stands self-condemned.

Thus at Kingstown, Jamaica, the dockyard and coaling depôt should be withdrawn from their present exposed position, and be retired to the inner harbour. Were this done, the general defence would be much less costly and yet stronger.

At some stations it is only necessary to provide for the security of the coal and the appliances used in coaling. For such a place the following simple method of defence has for some years been a favourite hobby of the writer's, and something of the kind has also been recommended by so high and experienced an authority as General Sir Lintorn Simmons, G.C.B., R.E. The scheme consists:

1. In stacking the coal at a distance from the water, and so situated that it could not be damaged by the guns of a hostile cruiser or flying squadron.

2. In connecting same with the harbour by a tramway, generally inclined so that the full trucks descending the incline would draw the empty trucks up.

3. In providing shoots similar to those used in Durham, Northumberland, and South Wales, for quickly loading barges at the end of the tramway.

4. In covering these shoots by an earthwork to protect them from hostile artillery fire.

5. In providing special barges so constructed that when scuttled they will just sink, and thus be hidden from a foe should he attack the harbour, and yet be easily recovered when he retires.

The method of coaling by means of barges is strongly advocated by many officers of the Royal Navy as preferable to all other means, and barges can certainly be loaded in less time at the shoots than they would take to unload at the ships.

One or two companies of infantry behind carefully constructed field-works would protect the coal depôt from any attack likely to be delivered on land, and the defence hardly requires a cannon or a mine. A few mines covered by quick-firing guns would add to the defence at no great expense, but such addition is not essential. The idea ruling such a defence is to place the objective—the coal—out of the reach of a cruiser. Any damage he can inflict on the shoots or the tramway could be repaired in a few hours, suitable material for repairs being kept in reserve at the depôt.

Even when it is desired to provide facilities for repairing defects in a ship's machinery or outfit, a great deal could be done at such a depôt, a fitting shop and store buildings being added; also a few strong trucks to carry loads of 10 or 15 tons, and a crane at the water edge to unload or load a barge with the special gear required. In short, the scheme is capable of expansion, and for distant stations in the Pacific has many advantages to recommend it.

Small Harbours to be denied to a Foe.—It is sometimes most important to deny certain small harbours to a foe, although during peace they may be of little or no commercial value. For instance, the little harbour of Balaklava was of immense strategic importance, and considering the large sums spent upon the sea forts and land defences of Sebastopol, the undefended state of Balaklava was evidently an oversight. Marsa Scirocco, near Valetta is another instance—and others could be cited.

Such harbours should evidently be dealt with so as to put difficulties in the way of attacking forces that might wish to utilise them.

The least expensive method would probably be a defence by purely automatic mines interspersed with mines under control from shore, where a few quick-firing guns would afford a certain protection to the mines and to the observing stations in connection with them.

Open Roadsteads and Coast Towns.—The defence of towns located, like Brighton, on the shores of the open sea has been much debated of late, and there cannot be a doubt that such places can only be effectively protected by the Navy. In a most excellent article in the *Times* of May 25, 1888, entitled "The Higher Policy of Defence," the writer truely said: "The command of certain waters exists when, within those waters, no hostile fleet can count on the time requisite for a serious enterprise without a strong probability of having a superior force to deal with." Thus, in a single sentence, the only true protection of our coast towns from attack by *a fleet* is clearly explained. There remains the attack by one or two swift cruisers. The great area of water puts mining out of the question, and shore batteries would be useless, for the bombarding range of a cruiser being, say, 7 or 8 miles, and the extreme effective range of guns on shore, firing at a small rapidly moving target, being, say, 3000 yards—it is evident that the cruiser could remain outside the zone of fire of the shore batteries and yet bombard the large target offered by a coast town, so that every shot would take effect. Evidently, therefore, the defence of such towns against such attack must be undertaken by guns afloat. Whether it is best to place these guns in swift vessels as recently recommended by Lord Armstrong, such vessels patrolling the coast; or whether it would not be preferable to mount them' on slower hulls, more heavily armoured, each stationed for its special work, is a matter for naval strategists to decide.

Conclusion.—In conclusion, as regards submarine mining, it is important to remember that each place will form a special problem, and that the plans for defence by sea mines should be drawn up by an adept well versed in harbour defence generally and submarine mining in particular. The artillery defence must be carefully noted, as well as the numerous local peculiarities of tidal current, depth of water, facilities for navigation, and other matters of this nature.

The object of any attack must always be kept in view when designing the mine defences, which should conform with the requirements of each situation.

CHAPTER XVI.

Boat and Steamer Equipment.

As regards the boats and steamers required for laying or raising the mines at any station, the establishment will vary according to the number of mines to be laid, the distances of the mine fields from the store depôts, and other considerations. At some harbours a small mooring steamer, say 45 ft. long, a junction box boat, and a few cutters and 12 ft. dinghies, are sufficient. At a large and important harbour where rough water may often be met with, larger mooring steamers are required, and an increased number of boats. Also, if the mine fields are distant from the depôts, one or more store vessels or lighters and a steam tug should be provided.

The Boats should all be strongly built, and should be fitted with smooth iron fair leads on the bow and stern, through which ropes can be led without chafing. Metal rowlocks that turn inboard when required to be out of the way of ropes, should also be used.

Junction Box Boats, or boats specially adapted for work connected with the junction boxes, should possess a bow joggle and small fore deck; and as it is desirable that the electrical connections should be made under shelter from rain or spray, the central well of the boat should be covered with canvas on a suitable framework.

A small hand crab is often useful. Although these junction box boats are generally towed into position and home again, it may sometimes be necessary for the crew to shift the position of the boat by rowing. Oars with rowlocks that turn inboard should therefore be provided.

The Steam Tugs may be any ordinary harbour steamer of moderate size used for this purpose. The small tugs on the Thames, Mersey, and Southampton water, would answer admirably—indeed some of them could with but little expense be fitted with special applicances and then act as mooring steamers.

The Store Lighters should be about 50 ft. by 15 ft., 4ft. draught and 3 ft. freeboard. They should be fitted with two steam or hand crabs with quick and slow speed, and two iron derricks having a sweep of about 15 ft. These derricks can be slewed by worm gearing driven by hand.

The lighters should possess ample room in the hold, and the central hatch should be under the sweep of the derricks, which may therefore be placed about 27 ft. apart.

The Small Mooring Steamers may be 45 ft. long, 10 ft. beam, 3 ft. 6 in. draught, and 4 ft. freeboard at bow. Such a vessel specially fitted for mooring sea mines has been designed by the writer and worked out in detail at Messrs. Day, Summers, and Co., of Southampton. She is light enough to be sent to any part of the world as deck freight in a large vessel, the boiler and machinery going separately. She is provided with a combination of derrick and winch, which has recently been patented. The derrick has a straight hollow mast of steel carrying a pulley at the top and another at the jib end for a wire rope with ball weight and hook. The mast pivots on a box fixed to the floor of hull, and the wire rope passes through the pivot centrally to a pulley secured to the hull, and thence to another under the winch which is conveniently fixed to the deck further aft. This is provided with two outside warping barrels driven preferably by steam on a secondary shafting, and a central drum driven independently by a worm and hand gear actuates the wire rope to derrick. The warping drums are employed for raising the mines, anchors, &c., to the surface, and the derrick for slinging them or bringing them inboard, after they come to the surface. The derrick mast passes through a strong collar secured to the deck, and immediately under this is keyed a wheel moved by a worm on a shaft leading to a position abaft the winch, where it is driven by a large handwheel that projects for nearly half its diameter through the deck. The difficulty encountered with the derricks hitherto employed, viz., that the weight on jib end had to be slightly raised when the derrick is slewed, and which has caused the proposed adoption of steam slewing gear in the British service, is thus obviated. Moreover, as the weights taken on the derrick are only raised for a few feet the combined winch meets the requirements in the simplest possible manner (see Figs. 88 and 89).

This small steamer is completely decked in (thus giving sleeping accommodation for the crew both in the fore and aft cabin), and the engine and boiler compartments are covered with suitable skylights. Good deck space is provided both at bow and stern, also the usual bow joggle, fair lead over stern, iron cleats and bulwarks, &c. She has three water-tight bulkheads, one in front of the derrick (a collision bulkhead), another in front of the boiler, and the third abaft the engines. The steering wheel is placed just abaft the latter bulkhead, and care is taken to give good turning power, viz., a circle of less than two lengths diameter. The engines, compound, non-condensing, but exhausting through a tank to heat the feed, drive a single screw propeller—a speed of about nine knots being obtained.

A Larger Steamer, 75 ft. long, 15 ft. beam, 4 ft. 6 in. draught, and 5 ft. 6 in. freeboard forward, with similar general arrangements, but with higher speed; a steering bridge, a small chart-room on deck, and altogether a more powerful boat and suited for rougher water, has also been designed in detail.

A second derrick can be placed on this steamer abaft the engine-room if desired, this being done now in our service; but, inasmuch as mines are slung to the sides of a mooring steamer by the cranes at a pier-head, or by the derricks of a store lighter, the necessity of such an addition is not evident. All mines and anchors should certainly be raised to the bow joggle.

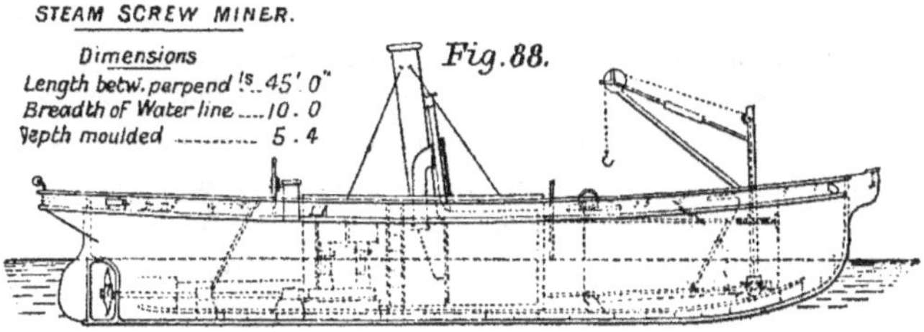

Fig. 88.

Fig. 89.

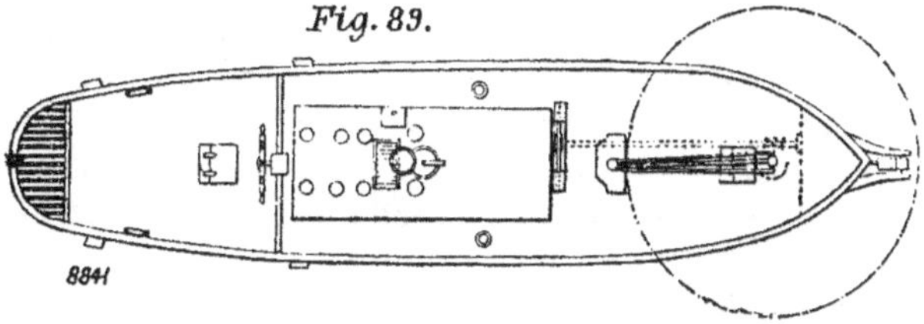

Such a steamer when fitted with sails can be sent to any part of the world. A vessel of this description would be a useful auxiliary in the active defence of a harbour after the mines are laid, and with this object in view fittings should be provided for mounting a Hotchkiss quick-firing gun, and an electric light—say a 60 centimetre Mangin lantern with inclined hand lamp. The dynamo and its engine could be fixed in the engine-room and be driven by the engineer for the propelling machinery. The master and deck hands of the steamers, and the crews of the smaller boats, should be well acquainted with the local peculiarities of the harbour in which the work is carried out, and they should be changed as seldom as circumstances permit.

CHAPTER XVII.

Practical Work.

Surveying the Mine Fields.—Some harbours have been carefully surveyed at recent date, and large scale charts exist showing the soundings with sufficient exactitude for sea mining; but many harbours have not been recently surveyed, and it is then necessary to check the depths of water at the positions of the electro-contact mines, and of mines fitted with circuit-closers, and of mechanical mines.

When the shore is not very distant, alignment posts can be set up showing the direction of each row of mines, or of junction boxes, and the angle therewith formed by a line joining one of the mine field points with a distant object being known by plotting same on the mine field chart, a boat can be placed in exact position by means of a sextant in the boat, or by means of a theodolite at the distant object on shore. A sounding can then be taken, and the time of doing so recorded. A tide gauge should be fixed at the pier-head, or otherwise, and readings taken every five minutes. The sounding can then be subsequently corrected for tidal level, and be brought to a common datum, generally mean low-water mark, spring tides. When the water is deep the soundings must be taken at slack tide, and a heavy lead, 15 lb. or 20 lb., be employed.

An experienced hand is required, or somewhat startling inaccuracies will be recorded and vouched for as absolutely correct. It has been suggested to employ instruments that automatically record the depth by means of water pressure indicators.

Sir William Thomson's apparatus arranged for depths up to 30 fathoms might be useful. Another instrument of a similar nature, but which at once records the depth by means of an electrical tell-tale on board, is now in the market. When using instruments of this description, it is not necessary to work during slack tide, and much time can therefore be saved by their use. In whatever way the soundings are taken, no trouble should be spared in order to insure accuracy, and thus avoid subsequent annoyance. Few things are more exasperating than to see a group of contact mines bobbing about on the surface at low water slack when it was intended to give them, say, 2 fathoms submersion at that time. Such a group would require to be raised and the error rectified when every one would be working at high pressure to get other groups down as quickly as possible. Having made a careful survey of each mine field, the length of the mooring lines can be corrected if any alterations in the soundings have occurred since the last survey. Check soundings should be taken annually to discover whether the depth at different places remained constant. This is especially necessary in the channels of a river and the entrance of harbours where the scour is considerable. At the mouths of some harbours sand banks are thrown up in a single night by strong gales.

Preparations for Laying Mines.—When it appears likely that the mines may have to be laid, the working parties should be organised in accordance with a previously prepared programme of operations, showing each day's and each night's work.

All machinery should be tried, coal and other stores issued, and every preparation made to start the work rapidly and systematically at a moment's notice.

Authority in the Queen's name should be obtained to prohibit fishermen and others from dredging and netting in certain waters, and to impose heavy penalties for disobedience. The friendly channels to be preserved through the waters to be mined should be buoyed, and a number of buoys that mean nothing should be laid to perplex a stranger and draw his attention from those which are important. Instructions should be issued to the men on the advanced lightships, if any, to extinguish the light in certain contingencies. Men told off to act as watchers should be sent to these vessels as they would be required for piloting and other purposes, such as cautioning vessels concerning any changes in the harbour navigation.

Laying the Cables and Mines.—Assuming that all the stores have been prepared, the first work to be done on receiving the order to lay the mines is to lay the trunk cables and to connect up the mines. These operations may usually be conducted simultaneously, and as soon as the first few groups are ready, they can be taken to the pier-head and slung to the mooring steamers, or placed on board the store lighters.

The more important mines, tactically, should be laid first, the advanced positions being of chief value in some harbours, the retired positions in others. If the tactical values of the various mine fields do not differ greatly, those in advance should be first attended to, for obvious reasons.

Laying the Trunk Cables.—As stated in a previous chapter, some of the main cables may often be laid permanently, and their ends buoyed.

It would not, however, be prudent to allow their positions to be known, and the buoys should therefore be treated like dormant mines (see Fig. 36, page 73). The explosive link may be arranged to hold down a small buoy attached to a line, and this to a chain by which the cable end and a small sinker can be raised to the surface whenever required. A heavy sinker should be attached to the cable at such a distance from its extremity that when the cable end is brought to the surface this sinker remains unmoved, and forms an anchor to which the boat employed in the operation can ride. A junction box may be added to the cable end as soon as the mines are ready to be laid.

Cables can be laid from an iron drum provided with a friction brake, or from a coil made on the deck of a steamer or lighter, or from the special barge already described on page 165. In laying from coils care must be taken to protect the men employed when clearing same, and it is always advisable to rig up some kind of brake to check the cable, especially when laying it in deep water. If the mine field be some distance from the shore, it is necessary to lay to the shore, and then land the shore end; but when the mines are near to the firing station the main cables can be laid from the shore outwards, the shore end being previously landed and connected electrically with the firing station by cables laid in deep trenches. When no main cables are used the cable from each mine is of course laid with the mine, and the shore end landed last.

Special shore end cables, with heavy armouring, which are used in exposed positions, should be laid permanently, and the other cables are then connected to the outer ends, thus avoiding landing operations when the mines have to be laid, and saving a good deal of time, even in fine weather.

Connecting up the Mines.—The operation of connecting up the mines with their cables, tripping chains, &c., can generally be performed in the open air on the connecting-up ground, but in wet weather it is desirable to do it under cover if a suitable shed be available. If not, movable jointing hoods should be used, as in the street work of the postal telegraphs. The depôt tramway should be led to the connecting-up ground, so that the mines may remain on the trucks from the time they leave the magazine until they are embarked.

The mines, loaded and fitted with their firing and other apparatus, having arrived on their trucks at the connecting-up ground, or shed, the crowned cable ends are fixed to the mouths of the mines by screw clamps, and the electric joint or joints carefully made. The tripping chains are then secured and stoppered to the cable with spun yarn, so that the cables are slack when a strain comes on the chains.

With buoyant mines the mooring lines are attached to the slings or attachment chains, but the sinkers can be connected to the tripping chains and mooring lines after the mines are slung on the side of the mooring steamer. A sinker can be put on each truck carrying a buoyant mine on its way to the pier-head.

There should be four or five trained men in each working party, and one to direct. With a little practice the men can be taught to connect up the different types of mines and the different arrangements with celerity, the men falling naturally into their places and performing each operation as required. In our service a party is told off by numbers, and each man's duties are detailed, but a hard-and-fast method of procedure for work of this character is not desirable, and causes a loss, rather than a gain of time. It is similar to riggers' work, and the method of procedure should be elastic rather than arbitrary. Hand sketches, with figured dimensions at length, are useful, and the officer in charge should supply them to the directors of the working parties whenever practicable.

Embarking and Laying Mines.—Electro-contact mines, in groups, on the fork system, can be embarked and laid singly if a store lighter be close at hand, but the mines for an entire group are generally slung to the mooring steamer, one mine with its sinker at the bow, the others with their sinkers on the sides of the steamer, and the branch cables arranged in coils on the deck ready and suitably for paying out. The slinging can be done by the crane at the pier-head, or by the derricks on the store lighter. Each sinker can be slung by a greased lowering line of sufficient strength, and of a minimum length of rather more than twice the depth of water where the mine is to be moored. Each mine is slung by a shorter lowering line. The mooring line is then shackled to the centre lug of the sinker, and the tripping chain to one of the side lugs. The mooring line and cable between the sinker and the mine are then coiled and tied to the side of the steamer with a piece of spun yarn that will break when a strain comes upon it.

Laying the Mines.—The vessel now steams close to the position of the first mine to be laid, which is found by alignment posts on shore, or by the sextant, or by both combined. The steamer is slowed and brought bow on to the current (if any), and the sinker is lowered away by word of command until the weight comes on the mine. The lowering line of the sinker is hauled in, and when the exact position of the mine is covered, the mine is lowered away also.

Another method which is more easily performed, but less accurate, is to sling both mine and sinker with short lines, and release them suddenly and simultaneously when the mine position is covered—care being taken that no sudden strain is thus thrown on the branch cable.

Whichever plan be pursued, the cable is taken aft and paid over the stern as soon as the vessel gathers way, and proceeds slowly to the group junction box, when a throw line is attached to the end of the cable, and it is transferred to the party in the junction-box boat, who proceed to make the necessary electric connections.

The other mines of the group are then laid in the same way. When a small steamer is used the mines should be slung so that they can be laid alternately from the port and starboard side, thus keeping the vessel in trim as far as possible.

When electro-contact mines are connected up in a string, see Fig. 60, page 125, their cables branching from the single main cable at a series of single connecting boxes, each fitted with a single disconnector, see Fig. 50, page 106, they are slung to the bow and sides of a steamer in a somewhat similar manner, but they are laid one after the other in one continuous operation, each being lowered as soon as the cable between it and the last mine laid becomes taut. The group being laid, the single main cable is paid out and carried either to a multiple main junction box (where it is connected to one of the cores), or to the firing station on shore. If a 75 ft. steamer be employed, an entire group may be slung on one side of the vessel, but with smaller steamers the mines should be slung and laid from opposite sides of the vessel in alternate pairs, in which event the cable must be passed under the mooring steamer from one side to the other when necessary.

Embarking and Laying Mines to be Fired by Observation.—These mines can be slung to the bow and sides of the mooring steamer in a similar manner to electro-contact mines arranged on fork, as already described. When each mine is lowered a man should be told off to stand close to the lowering line and lower a flag or make other visual signal. Observers on shore should at the same moment bring the collimation of the observing instruments on the position of the flag, clamp them when the flag is lowered, and keep a record of the angles as shown by the instruments.

Mine Field Repairs.—When the electric testing shows that a group of electro-contact mines is out of order, a suitable boat should be sent to the group box, the buoy, if dormant, being first brought to the surface by firing the explosive link. (If the buoy will not rise the defect probably exists between shore and box, and the cable must be underrun by a steamer.) Should the main cable be in good order, the box is raised, and each mine tested singly, the faulty mine or branch being thereby discovered. The branch cable being disconnected from the box, is then transferred to a mooring steamer by means of a throw line, and the branch cable is underrun and carefully examined as it comes inboard. As soon as the tripping chain comes to the surface it is taken to the bow joggle of the steamer and carried to the warping drum of the winch or capstan—preferably driven by steam. The sinker is then raised, the mine secured as soon as it comes to the surface, and the fault discovered at once if possible. Failing this, the mine is taken to the store ship or to the depôt, where it is disconnected and the fault localised. If the necessary repair can be quickly rectified the mine is relaid, but it generally saves time to have a small reserve of loaded mines ready, and to connect up one of these and lay it at once.

When the electric tests show that an observation mine is out of order, a suitable boat should be sent to the multiple main junction box to raise it. The mine cable is then transferred to a mooring steamer, which should underrun the cable and raise the mine, the entire operation being very similar to the one just described.

Should a group of electro-contact mines moored in series get out of order, its repair is a serious matter, as the whole of the group must be raised, even if the fault be discovered in the first mine. The mines are nearly certain to be dragged out of position directly the steamer raises the first mine. It may, therefore, sometimes be preferable to try the curative effect frequently obtainable by applying the firing current, already alluded to in a previous chapter.

Mines are moored in so many different manners—some with and others without detached circuit-closers, some on single and others on double moorings, some on the bottom and others buoyant—that it would weary the reader if the laying and raising of each were given in detail.

CHAPTER XVIII.

The Personnel and the Stores.

The organization and superintendence and designing required for submarine mining are performed in our service by the officers of the Royal Engineers, who are frequently expert boatmen, and who, from their high educational attainments, are easily trained in the scientific portions of the work. It is rather hard upon them, for experience shows that it practically destroys their chance of seeing active service in the field. As for the men it is impossible to give any sound reasons for employing our most expensive and, under many conditions, our most useful soldiery in this manner. Our small force of Royal Engineers would be urgently required at the front in the event of a war of such magnitude that our harbours would have to be mined. The same argument applies with equal or greater weight to our Navy, and to a less extent to any other portions of our fighting forces.

For this reason, if for no other, it is important that civilians should perform all the services possible which are connected with harbour defences. Submarine mining is certainly one of them. An attempt is being made at some ports to employ the Volunteers as miners, and no pains should be spared to insure success if possible, but the results to date are not encouraging. It is not a popular service with them. It is rather cold, wet, and dirty work with very little soldiering about it. Much of it is very like oyster dredging with more tallow, tar, and twine. Only a small proportion of the men are required for the scientific arrangements connected with the electrical gear. After an experience extending over a number of years I am convinced that most of the work can be properly and economically performed by civilians under the superintendence of trained officers. At Singapore and Hong-Kong, Malays and Chinese boatmen, who cannot speak the English language, and who are only instructed for short periods annually, have been employed with satisfactory results; the work being directed by a small nucleus composed of highly-trained regulars. Much more then should civilians and volunteers at home be able to act in a similar manner, but the nucleus should be composed of men permanently employed at each port.

Intelligent men of experience in boat or steamer work, soon learn how to proceed. To fill a drill-book with directions that No. 1 shall do this, No. 2 that, and so forth, savours of military pedantry. The civilians employed by the dockyards or by the Trinity Board in laying and raising channel buoys are not drilled by numbers, and in practical sea mining the drills are seldom if ever followed. Indeed, it is not possible, because the circumstances vary so frequently. As a rule, one or two good men in each squad do most of the work, and the rest assist when required. No doubt it is difficult to convert soldiers into sailors; but the drill-book assists in the process no more than the guernseys and men-o'-war caps that are donned.

Whenever possible the men employed for the water work connected with submarine mining should be able-bodied seamen, who should sign articles and be under the discipline of the masters of the mooring steamers, who themselves are under the orders and guidance of the superintending officers.

The rough work on shore, such as coiling cables, moving weights, &c., can be done by labourers, and the electrical work can be done by permanent hands who have received a special training, whether it be in or out of the service. There is not the slightest necessity for a single scarlet jacket or pipe-clayed belt. The men should be clad as boatmen. They should not be moved *en bloc* from one station to another at the beck and call of the adjutant-general. They should be or become acquainted with the local peculiarities of the waters of their port. The working parties and the programme of operations for the day should not be upset by unforeseen regimental troubles. If men misbehave themselves as civilians they can be discharged. A number of men would not daily be required for all kinds of regimental and garrison duties. In short the work during peace would be done, where it is now done at all, with a considerable reduction in the numbers of the personnel, and a still greater saving in £ s. d. The idea that Sapper labour is cheaper than civilian labour is entirely fallacious. In a paper read at the Royal United Service Institution on March 18, 1887, I showed that the cost of a high-class civilian crew in a large mooring steamer came to 5d. per working hour per man, and that the cost per working hour per man of a submarine mining company of Royal Engineers was 1s. Good labourers can be hired for 3d. in the winter and 4d. in the summer for temporary employment, and expert boatmen for 6d. an hour. But efficiency is of more importance than cost, and if it could be shown that the service is better performed by Sappers than it can be by civilians no one would desire a change if the Sappers were not wanted in war for other duties. But this cannot be shown. The companies are frequently moved, the men transferred before they learn local peculiarities, and the works are interfered with by regimental necessities.

Those who advocate soldiers for sea mining fear that civilians would desert when war came upon us. If so, they must also anticipate the desertion of the personnel from Her Majesty's dockyards, as these men are as likely to be "under fire" as the submarine miners. Higher pay would be expected and probably be given, but no man worth retaining would desert at such a time. The personnel for an ordinary station might be somewhat as follows: Two or three highly trained officers for directing; one mooring steamer with civilian crew; one mooring party, six civilians, under a good leading hand; one depôt party ditto; one storekeeper. Engine drivers and electricians, according to requirements, to keep the machinery and electrical equipment in good order. Artisans, labourers, and boatmen hired as required. Also one small company of volunteers under local officers who are professionally connected with work of a similar nature; the employers of labour in small shipyards, for instance. Such a company should be recruited from the local boatmen, shipwrights, and artisans. Men should be encouraged to join, not boys. Directly war appeared to be imminent, additional civilian electricians, engine drivers, and labourers would be secured. Much difficulty will always be met with in finding time and opportunity to train each volunteer in the multifarious duties of a submarine miner. It is therefore desirable to divide the work as far as possible, and teach the men in squads to do a few things well, rather than the whole indifferently. If experience should prove that the volunteer movement is not applicable to this class of work, an honest attempt should be made at some of the stations to work with a permanent establishment of civilians, and a comparison could then be made between the relative cost and efficiency of this and the present military organizations.

The Royal Engineer and auxiliary forces now employed are quite insufficient, and the danger caused by this deficiency is increased by the extreme and totally unnecessary complexity of our service arrangements, which entail a long and careful training of the personnel in order to insure success.

This training should in any case be simplified as much as possible by teaching the men certain duties and keeping them to these duties, instead of making them so many jacks-of-all-trades and masters of none. If a man is to drive an engine when war is declared, why teach him electricity? If another is to operate an electric light, why teach him submarine mining? The officers alone need to know it all thoroughly.

As to the rest, there should be division of labour and division of instruction. With a military system it is very difficult to do this. Tom goes on guard and John must take his place on the works, or the works must stop. Each must know a little about a great deal. It is far better for them to know a great deal about a little. With civilians on the works this idea can be followed out. But whatever may be the personnel, and whatever the system of firing and testing the mines which is adopted by any country, they should be such that the mines can be very quickly laid after the order to do so has been received. If our present system and personnel be considered satisfactory, let the adjutant-general suddenly and unexpectedly give an order that the mines shall be laid forthwith in all our harbours which are assumed to be prepared for such an event, and let him countermand the order at the end of a fortnight and ask for reports from all these stations showing the progress of the work at that time, the number of mines laid and in working order with their firing stations and operators, the number of test rooms fitted and provided with the necessary trained testers, the number of electric lights efficient, &c. The results would perhaps open the eyes of the authorities to the fact that secrecy in submarine mining is now made with the indefensible object of hiding our weakness, not our strength.

The position and number of the mines only need be secret. All the rest should be brought to the light of day, and until this is done and public criticism can be brought to bear upon it, the present complicated system and utter want of proper means to work it, and the present lack of organization will continue. Some fine day a great naval war will throw us on our beam-ends, and the officers at every station will cry out in vain for the highly trained electricians and personnel generally which are required. A trial at one station, like that which took place at Milford Haven, is no test whatever, because personnel, stores, and steamers were promptly transferred from other stations to the scene of operations. Even then it took weeks instead of days to get *a section* of the defences into working order.

Our mines contain apparatus requiring a number of delicate and difficult adjustments which are easily put out of order after the mines are laid. The mooring steamers then have a lively time, daily repairs for some or other of the mines being generally necessary.

A foreign officer told me recently that some mines had been laid in his country for five years and are still in good order. I do not remember any instance in which our mines remained down for five months, only a portion will remain in good order for as many weeks, and some are unserviceable directly they are laid.

Our system is so intricate that success cannot possibly be secured with any certainty, and the instruction of efficient operators is most difficult and tedious.

These remarks apply not only to the mines, but to the apparatus on shore, especially in what are termed the test rooms. Here will be found a perfect network of wires, commingled with galvanometers, telephones, commutators, batteries, wandering leads, &c., on which various electrical tricks can be performed. If all this delicate apparatus could keep the mines in an efficient state, something might be said for it. But it has an opposite tendency, and although it is absolutely necessary to test mines arranged on our service systems, because so many require renewal or repair after brief periods, it would nevertheless be infinitely preferable to employ strong and simple gear, not easily deranged, and therefore requiring but little if any testing.

The same love for the complex permeates the larger stores. Steam machinery is employed where hand gear is preferable, as being less likely to break down at a critical time, or to be damaged by a chance shot. For instance, the mooring steamers contain much unnecessary machinery, steam cranes are used on shore where hand cranes would be better, and so on. If doubt be placed on these opinions, let a civilian electrical engineer of high standing be appointed to report on our mine systems; and a good mechanical engineer on the hoisting and mooring and mine raising arrangements. Important simplifications would soon be insisted upon, and the question as to the personnel would then become proportionally easier to solve.

The Stores.—Let us now consider how the stores should be procured. The Italians have a large government manufactory for submarine mining stores, and for some years these stores were largely manufactured for our service by the Royal Laboratory at the Royal Arsenal. Experience, however, proved that this system was costly, and that vexatious delays frequently occurred, this extra work being more or less shelved when it interfered with the legitimate work of the manufactory. The requirements for submarine mining come in fits and starts, and are not well suited for providing continuous work in a special department. It would, therefore, have been bad policy to start a government factory for their special manufacture; and there is no necessity to do so, for all the gear, except the instruments, can be readily made in good engineering shops.

Moreover, the resources of the Royal Arsenal would, in the event of war, be taxed to the utmost, and the speedy manufacture of submarine mining stores could not then be relied upon. For these reasons the resources of private firms have been utilised in this country, and with considerable success.

A few good firms were selected by the Royal Engineers, and the orders were for some years placed with them, often without competition, when the prices quoted were considered to be fair and reasonable. In the year 1878 large quantities of stores were obtained very promptly in this manner, the Royal Engineers directing both the purchase and inspection. The ordinary War Office system was thrown to the winds, as it always must be in times of emergency. It is never applicable to scientific stores. The system of lowest tender is pernicious when applied to any stores of the kind. It is alike inapplicable to artillery and to engineer stores, except perhaps for such articles as picks and shovels, when a rigid inspection and high specification and pattern may perhaps protect us sufficiently.

The recent disclosures before the Committee on the Sweating System have shown that even with such articles as tunics, belts, and saddlery, the present system of contracting with manufacturers at the lowest price in an open market fails. Much more then must it fail when it is necessary to obtain scientific instruments and stores. The Contract branch is perfectly satisfied so long as the stores just pass inspection. It frequently happens that a large percentage are rejected on delivery, causing a serious delay, which might have the most serious consequences were we likely to require them in a hurry at an outbreak of hostilities. The present system produces prices which prohibit high-class work, and firms that pride themselves on never turning out any other kind of work cannot secure an order. Shoddy reigns supreme, and laughs in his sleeve at an officialdom which he endeavours to hoodwink at every turn, generally with facility, as instanced by the two-ply thread, &c. It is no longer an honour, as it used to be, for a firm to be on the Government list of contractors.

People absolutely unknown as instrument makers are asked to tender for the manufacture of confidential instruments that require the greatest nicety of adjustment.

Tenders for steamers are advertised for ! &c., &c.

The duties of the inspecting officers have become unnecessarily difficult and arduous.

This system strikes at the root of all good work. It encourages middlemen, subletting, and sweating. The middlemen employ indifferent workmen, often men who have fallen from a good position, and who could not obtain a place in any respectable manufactory. These men are ground down to work for wages out of which it is almost impossible for them to house and feed themselves. Their families are either starving or in the workhouse. The legitimate manufacturer has either to discharge honest hands or pay them such wages that he is almost ashamed to negotiate with them. The sweating system is gradually permeating the whole of English trade. It is not confined to boots and belts. It is bringing starvation to the doors of our working millions, to our political voters. The present system of Government contracts in all branches encourages this undesirable state of affairs.

The manufacture at the Royal Arsenal, with all its delay, was far better than this. The stores were excellent, if costly. The remedy is evident. A return should be made to the system followed in former years. The Government lists of contractors should consist only of well-known firms celebrated for honest high-class work, and the users of the stores should have a voice when the lists are drawn up.

Competition between firms well known for sound work is sufficiently keen to keep prices down to a proper level for such work, and if they only competed among themselves, and had not to compete with the sweater, the prices would be such that a healthy emulation to produce the best article would be possible.

Stores for submarine mining, like those for torpedo gear and artillery, cannot be made too carefully. So much depends upon each link of the chain between mine and firing station being in good order. A leaky rivet, a weak shackle, a bad electrical connection, or any one of a hundred little matters of the kind, may destroy the efficiency of an important mine. Cheap gear is a mistake, because it is likely to fail, and may do so at a critical moment, when perhaps the destruction of an enemy's ironclad is at stake. Now the cost of such a vessel would mine twenty or thirty harbours, and her value in war might greatly exceed her cost.

CHAPTER XIX.

AUTOMATIC MINES.

ENGLAND cannot afford to put impediments in the way of her own commerce in time of war. Any serious stoppage to her ocean trade would be almost tantamount to defeat. Automatic mines can, therefore, only be used on our defences to a limited extent, and this has been recognised from the first. Foreign nations are not in the same position, and many of them have devoted much time and labour in order to perfect this form of mine. For instance, where a relatively weak naval power has to quickly block its rivers and harbours on the outbreak of war, the elaborate electrical arrangements adopted in our service could not be prepared in time. Moreover, its commerce on the high seas would certainly be stopped, and its harbours be subject to partial if not absolute blockade. Under such conditions good automatic mines are both useful and necessary. But a strong naval power would require them as a naval arm, to assist in blocking the harbours of a foe, by sowing the entrances with mines of this nature. Even in the harbour and river defences of England, there are some situations where automatic mines may advantageously be laid, the shore being so far off that controlled mines could not be worked satisfactorily. It is therefore to be regretted that a good pattern has not been perfected by one or other of our services. Too little attention has been paid to the subject. Improvised mines are not good enough. On the other hand, the mines in store for defence purposes contain too elaborate an apparatus, and, in any event, they cannot easily be converted into automatic mines for blockading purposes. No attempt has been made to do so, and an Admiral who recently stated in the *Times* that we are provided with the mines for such operations, must have been misinformed.

An inherent defect of most automatic mines is their vulnerability to countermining. For this reason the cases need not be strong. If they can resist the blow of a countermine which will just cause the mine to explode by shock, it is enough.

The chief requirements of an automatic mine are that it should be quickly, easily, and safely laid; also that it should be safely recovered when desired. Wave action should not cause self-destruction, and a mine should become automatically safe if it break away from its fastenings. Moreover, as it would manifestly be impossible for vessels to accurately survey the waters and lay the mines at the entrance of a harbour held by an active foe, such mines should be provided with an arrangement for bringing them automatically to the required submersion, the mooring line being unwound until the proper length is obtained.

Scores of inventors have attempted to produce a good automatic mine, but hitherto without complete success. A few of the different types will now be examined.

Chemical Automatic Mines.—Professor Jacobi's torpedoes were used in the Baltic by the Russians during the Crimean War. These mines were fitted with a glass tube containing sulphuric acid, the tube being imbedded in a mixture of chlorate of potash and sugar. Levers projected from the mine, and were so arranged that a passing vessel on striking one of them, caused the glass tube to be broken and the charge of gunpowder to be ignited. Defects: Action slow; charge employed too small; dangerous to lay. Dangerous to recover or destroy when no longer required. Results during the war practically *nil.*

About ten years later the Dutch Government carried out experiments with mines primed with Colonel Ramstedt's exploder, viz., a bent glass tube containing a small plug of potassium covered with naphtha. The long leg of the tube is sealed, and the short leg open, but it should be covered with a thin skin diaphragm. This leg is in contact with the priming charge. The long leg is carried through the case by a waterproof joint, and is exposed to the sea water. This can be covered with a perforated leaden cap which is bent by a vessel striking it, thus causing the tube to break, and water to get at the potassium, igniting the charge. Defects: Action rather slow; dangerous to lay; does not remain in good order for lengthened periods. Dangerous to recover or destroy when no longer required.

An improvement, suggested by the writer, consisted in surrounding the projecting glass tube with a perforated brass cylinder containing a small weight with a vertical hole fitting over the tube; also, in filling the cylinder with a cement slowly soluble in water, such as a heated mixture of sugar and chalk. This makes the mines safe to lay, as the apparatus only becomes active after the period necessary for dissolving the cement. Moreover, the charge can be so placed in the case that the mine turns over if it escapes from its fastenings, the weight tumbles out of the cylinder, the latter protects the glass tube, and the mine can then be recovered with safety. Defects: Action slow, and mine does not remain in good order for a long period.

Frictional Automatic Mines.—The mine invented by Mr. E. C. Singer and extensively used with much success by the Confederates in the American War of Secession, is a good example of this type. A friction tube and small priming charge is secured to the bottom of the case, and is suitably protected from the salt water by a thin diaphragm. One end of a short chain is attached to the pull ring, the other to a cast-iron cover resting on the top of the case. A central aperture in the lid surrounds the ring for the lowering line. Until the latter is withdrawn the mine is consequently safe. As additional security a short bight of the chain is connected with a safety pin secured to the bottom of the case or to the mooring chain. The last operation is to withdraw this pin by a line, thus making the mine active. This can be done from a safe distance. The line can be secured to a float and the moored mine remain passive as long as desired. If, however, the lid has been thrown off in the mean time, the mine would explode on withdrawing the safety pin. Defects: Dangerous to recover; liable to self-destruction by tilting in a strong current. Easily destroyed by sweeping operations.

Percussive Automatic Mines.—Several patterns of this type have been invented, and tried experimentally in this country; but it is so difficult to insure safety when recovering the mines that none of these patterns have been adopted. Most of these apparatus are secret, but one of them which was invented by the writer can be described, and forms a good illustration of the type. The bottom of the mine case D is provided with a circular hole to which the apparatus is secured by bolts and nuts. The apparatus consists of a cast-metal hat-shaped chamber A, smaller at the top than the bottom, and carrying a tube with a helical spring *h*, that actuates a striker, on a cap *c*, which ignites a detonator and the priming charge in P. The bottom of the chamber is closed with an india-rubber diaphragm *i*, carried on a central bolt *x* by a suitable nut and washer. The bottom of *x* passes through a lower casting B of peculiar shape (see sketch, Fig. 90), and containing in its lower portion some soluble cement, such as a mixture of powdered chalk and burnt sugar. This chamber Z has orifices communicating with the salt water outside. The mine is moored by the ring R. Into the top of *x* is screwed a small spindle of brittle steel, with an eye at the top, by which a cord connects it with the striker. A weight W rests on the nut. This weight cannot rock in the upper portion of A when in the position seen in sketch, but when it is pulled down into the lower portion its movement horizontally is only checked by the fragile steel spindle.

When the mine is first laid its buoyancy acting against the mooring cannot move the rod *x*; but after a few hours the cement in Z dissolves, the rod is pulled down, the spiral spring is compressed, and the weight W is brought into the larger portion of chamber A. If a vessel now strike the mine the inertia of W breaks the steel spindle, and the striker is released upwards, exploding the mine. Thus, these mines are quite safe to lay out. The difficulty is to raise them again when required.

The plan proposed by the inventor was to employ a small explosive link at the foot of the mooring line, the link being exploded by electricity, and to carry a small insulated wire to a position of safety, say 100

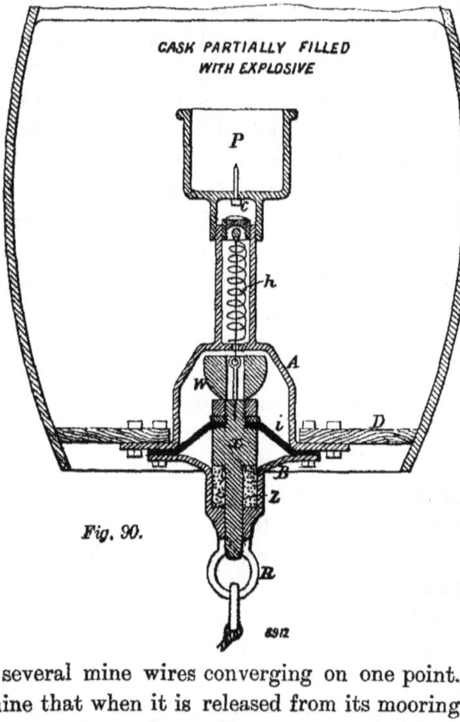

Fig. 90.

yards away, several mine wires converging on one point. Also to so ballast the mine that when it is released from its mooring it shall turn over as it floats to the surface. The pressure of the water acting on the rubber diaphragm and the weight of W then bring the apparatus into the safe state which was obtained when it was first laid. The position of R can be seen when the mine floats up. If home, the mine is safe. If not, the case should be destroyed by rifle bullets, and the mine sunk, an explosive being used that is damaged by water. Other means than electricity can be used for releasing the mines from their moorings. Defect: Difficult to recover with absolute certainty as to safety.

Electric Automatic Mines.—This is doubtless the best and most reliable type of the automatic mine. It has been adopted by several of the Continental powers in one form or another, a favourite pattern being the invention of Professor Hirsch. In this, each mine is provided with a number of projections or nipples, against any of which a boat or vessel may strike. Each nipple consists of a leaden envelope containing a small glass phial full of acid. When this is broken it is arranged that the acid shall run down upon two electrodes, thus forming a voltaic couple. Wires from the holes in each nipple are carried to the mine fuze, which is consequently exploded directly one of the nipples is bent. Part of the circuit is arranged to pass through a loop made by two wires led from the mine to a safe distance, and these are connected after the mine is laid, and when it is desired to make it active. Until this has been done the mine cannot be exploded, but it can be damaged if struck. In order to raise such a mine, it is only necessary to recover the loop, and to disconnect the wires. The mine is then safe to raise. Of course there is danger that a faulty connection may exist inside the mine, and it would be more assuring to have the source of electricity outside it.

This leads us to another and probably the best pattern of the electrical type of automatic mine. It was first suggested by Sergeant-Major Mathieson, of the Royal Engineers, and consists of the usual electro-contact mine connected with a voltaic battery in a suitable water-tight box which is submerged near to the mine, but far enough from it to allow of the battery being raised and disconnected with safety before raising the mine.

A number of mines can be fired in this manner from one battery, and the recovery of the latter when desired can be easily insured by several obvious devices without using a marking buoy.

The only difficulty in connection with this arrangement has been the constancy of the submerged firing battery, which is apt to deteriorate if the water-tight chamber be small.

The very sensitive low-resistance detonators that have been introduced by the Danes (see page 117) has simplified this problem immensely, as a small number of cells can be used, and the cells themselves can be of moderate size. A pattern should be employed that is known by experiment to remain in working order for lengthened periods without attention, and means must be taken to prevent loss of liquid by evaporation or otherwise.

The same mines and gear generally can be employed for this system as for the electro-contact mines controlled from a distance, but, as already stated, it is unnecessary to provide strong and costly cases for automatic mines of any pattern, because their self-destruction is so readily brought about by countermining.

Electrical automatic mines arranged in groups, each group having a submerged voltaic firing battery, are doubtless the best for harbour defence, their only defect being vulnerability to countermining. For naval purposes, however, the mines could be laid more rapidly and conveniently if each were complete in itself, and one of the other systems should perhaps be chosen. Moreover, for naval purposes, such as blocking the waters of a foe by mines, it is advantageous to use some arrangement whereby each mine will take the desired submersion automatically when it is laid. Such an apparatus has been invented by an officer in the Royal Navy, but it is a secret, and cannot therefore be described on these pages.

A good pattern of automatic mine is much needed for our harbour defences, and those stations that require them should be supplied. For instance, important stations partly surrounded by coral reefs, pierced in several places by navigable guts or passages too far from the land for the electrically controlled systems, should no longer remain unprovided with automatic mines of a simple, strong pattern, not likely to get out of order or to require repairs. Surely a little time and attention can be devoted to this important subject by those concerned.

"*Frame Torpedoes.*"—When shallow waters have to be mined, and when the current always flows in one direction, as in the large American rivers, a very useful and efficient form of mine is the frame torpedo employed by the Confederates in the War of Secession. A framework of strong balks is made and a mine fixed on the upper end of each timber. The lower end of the frame is secured by short chains to several sinkers, and the upper end is held down to any desired submersion by a chain or chains to a smaller sinker or sinkers. The lower end of the frame is up stream, and consequently the torpedoes are presented to any vessel advancing up the stream. Various devices can be used for ignition on contact, the best probably being some arrangement whereby a plunger is forced in by contact with the vessel, and thereby releases the striker of a lock action inside the torpedo.

The size of the torpedo can of course be adjusted so as to carry any desired charge, and to possess any required buoyancy, but the latter may be avoided if the framework possesses sufficient buoyancy, due allowance being made for water-logging. In swift currents the buoyancy must naturally be greater than in sluggish streams. Chains should be laid up stream from the frames, so that they may be recovered with as little danger as possible by vessels towing them up stream to some spot suitable for dismantling them.

Similar mines can be fixed to piles.

CHAPTER XX.

The Attack on and Defence of Mined Waters.

The defence of mined waters must usually be effected by a powerful artillery fire, but the guns must be of the proper quality for their targets.

The necessity of mounting heavy armour-piercing guns is universally accepted, as evidenced by the costly forts and batteries which have been erected throughout the world wherever harbour defence has received attention.

But an attack on mined waters must be delivered by small craft; and, as it is not desirable to crack nuts with steam hammers, the artillery to ward off such attacks should include a large number of quick-firing guns, such as the Hotchkiss or the Nordenfelt. These guns should be so placed on shore that they are not inconvenienced by the smoke from guns of heavier calibre, and should be so mounted as to remain under cover until the moment they are required. Moreover, smokeless powder should be used in their ammunition (see page 224). But boat attacks should often be met afloat, especially where the mined waters are some distance from the shore. In such event, quick-firing guns should be mounted on a number of small steamers, this defence flotilla being kept out of the range of the attacking artillery as long as possible.

Steam yachts and harbour tugs would answer well, and a number of quick-firing guns should be kept in store for their armament.

Assuming that the defenders possess torpedo boats, these should not be employed in repelling boat attacks. Unless they are nearly submerged and of the almost invisible type, they should be held in reserve to act against large vessels during the later stages of the operations when the defenders' artillery has been crippled, and the waters have been partially cleared of mines. At such a time a reserve of defensive power would be of great value. As attacks upon mined waters are probable at night, the defence must be well supplied with powerful electric arc lights to search for and discover the positions of the attacking flotilla and of the larger vessels in support. Those lights which are used principally in connection with the artillery should be under the charge and control of the battery or fort commander. Those again which are provided specially to aid in firing the mines at the correct moment, should be under the control of the officer in charge of the mines. When an attack has been developed, the quick-firing guns of the foe might soon destroy the electric lights on shore, and means should therefore be taken to use plane mirrors as reflectors when the foe comes to close quarters. Thus, when the lights are used for searching purposes, the costly catoptric, or dioptric, or catadioptric lantern can be exposed ; but when the light is used for illuminating the waters near at hand, the lamp should be lowered under cover, and the plane mirror raised.

The small steamers already mentioned would act on outpost duties in advance of the mined waters, and the search lights would greatly assist them. Some of the larger patrolling steamers might with advantage themselves carry electric lights, say, 60 cm. Mangin lanterns containing inclined hand lamps and arc lights, each possessing a power of about 20,000 standard candles.

The first step taken by the attack would be a heavy artillery fire, delivered on those portions of the defence which are within range and exposed to fire. This would be done by daylight. To attempt any boat attack before the artillery defences had been crippled as far as possible in this manner, would invite defeat. The patrolling boats could not hope to offer an effective resistance in the advanced zone, except for a short time. They would retire as soon as the attacking flotilla had reached the waters within range of the quick-firing guns on shore. Good range-finders would be most useful at this stage, as also efficient night sights for the quick-firing guns. An attack, which must always be in the nature of a forlorn hope, may then be repelled before any serious damage is done to the mines or their gear.

Actual war can alone decide whether such attacks can ever succeed against an energetic and well-conducted defence. Operations of this nature undertaken in time of peace are probably more misleading than instructive. They leave so much to the imagination of the umpires.

Creeping for Cables and destroying them by means of explosive grapnels fired by electricity is a trustworthy method of attack, and as it can be undertaken by small row-boats, the element of surprise assists the operation on dark nights when boats cannot readily be discovered. In order to throw difficulties in the way of such a method of attack, the waters in front of the mined areas should be liberally sown with pieces of old cable, lengths of chain, &c. ; and these bottom obstructions become more efficient if they are connected with small sinkers here and there, and with submerged buoys between the sinkers, the loops of dummy cable or of chain thus formed near the bottom being designed to catch the grapnels.

Sweeping for Mines by boats in pairs, with drift ropes or nets between them, is a most difficult operation to carry out at night, and it would of course be impossible to perform it successfully by day under a hostile fire.

The Attack by Countermines finds many advocates, and its efficiency is fully believed in by a number of naval officers whose opinions on such a subject should carry weight. Nevertheless it cannot be denied that the methods usually adopted in Europe for laying countermines are crude, and that any of a number of small accidents which are nearly certain to occur on active service would cause the failure of a boatload of countermines. To lay a string of mines from a boat is a most difficult operation at the best of times, in full daylight, and to do so successfully at night, during the excitement, smoke, and turmoil of an actual engagement, when a single bullet may cut one of the exposed electric cables, would be scarcely less than a miracle. The

Fig. 91.

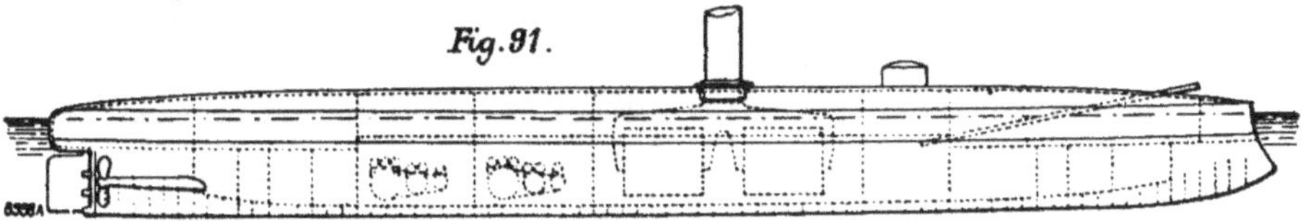

Americans are developing a system which promises to act much better. It is proposed to employ the air gun now being perfected by Captain Zalinski, United States artillery, and to mount three of them on a steamer specially adapted for countermining and other siege purposes (Fig. 91). This vessel is designed to have a moderate speed, a draught of 15 ft., good beam, and low freeboard. Her displacement to be about 3500 tons, and her under-water hull to possess an extreme cellular subdivision, and a double bottom 18 in. deep, extending to a height of 2 ft. above the water line. This double bottom to be filled with cocoa fibre cellulose, which quickly swells and fills any shot-hole through which water might otherwise enter. Her top sides to be armoured to 4 ft. below the water line, and her deck to be turtle-backed and covered with 5 in. of steel, the armouring weighing a little over 1000 tons in all. Her three air guns to be capable of throwing shell containing 100 lb. charges of high explosive to ranges of at least two miles, and larger charges to shorter distances.

The designer considers that 100 lb. of blasting gelatine will destroy all mines within a radius of 50 ft. He proposes that three air guns should be mounted side by side in the fore part of the vessel, the two outer guns being so arranged that they can be traversed a few degrees outside the central line of direction.

It is intended that the vessel shall anchor outside the waters supposed to be mined and take up any required position for the operation of countermining. The range being settled, the outer guns can then have their traversing gear so adjusted that their shells shall drop 100 ft. on either side of the shell from the central gun. A volley being fired, the range is altered by 100 ft., and another volley delivered. The traversing of the outer guns is altered when required, so as to retain the same width of countermined channel at all the ranges. In this way a channel 100 yards wide and two miles long can be countermined in two hours without shifting the vessel, 160 shell being expended per mile. If desired, shells specially fitted with releasable buoys can occasionally be discharged from the outer guns, and the channel be thereby buoyed as it is cleared. About 20 tons of ammunition, costing 6000*l.*, would be expended per mile of cleared channel.

Captain Zalinski states that the range is practically unaffected by the slight alterations in the elevation of the guns, which would be caused by the vessel pitching at her moorings. The vessel would carry as additional armament a number of quick-firing cannon and machine guns.

It is evident that countermining operations, even when undertaken from a distance by such a vessel, cannot be successfully performed by daylight so long as the artillery defences remain; and considerable uncertainty would attend the operation when conducted under cover of darkness, however carefully the ship is laid by compass bearings or by lights on shore, should they exist. Sweeping and creeping could never be undertaken by daylight, within range of the shore, so long as the defence retained the power of bringing machine gun fire to bear upon the boats. In short, the whole of the work connected with the attack on mined waters is so difficult and hazardous that we may fairly doubt the probability of these operations ever being seriously undertaken until the forts and batteries have been demolished and the defence thoroughly demoralised. Nothing of the kind occurred during the American War of Secession, when the vessels either ignored the mines and took their chance, or attacked the stations by forces landed for this purpose, or remained outside; and of these three courses, the latter was generally pursued.

Illumination of Fortified Harbours.—Nevertheless, as the attack of mined waters at night is so favourably viewed by many experts, the defence must be arranged so as to offer a strong resistance at that time, and the illumination of the mined waters and of the waters in advance of the mines becomes important.

The French have devoted much attention to this subject, and have,

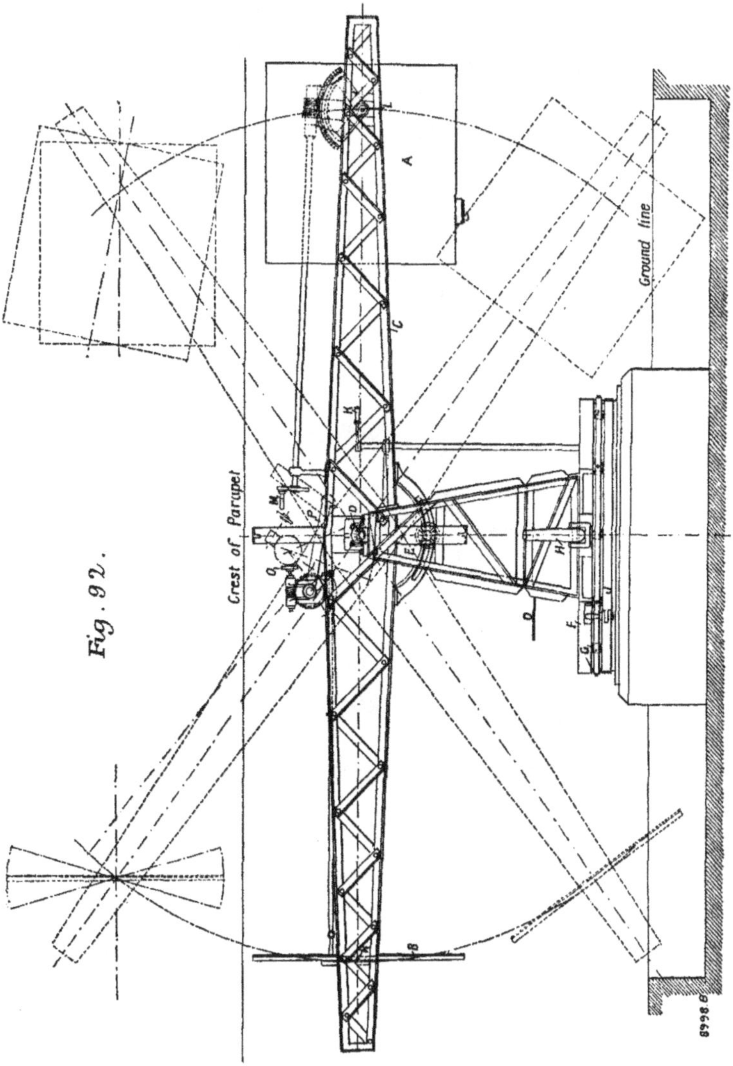

Fig. 92.

it is believed, erected a number of powerful electric lights at each of their sea fortresses. We have done the same, but in a less pronounced manner; and when we consider the great cost of these lights, their extreme vulnerability to machine and other fire, and the inconvenience to the defence flotilla which their injudicious use may often occasion, it would appear desirable to limit the electric lights to a moderate number at each harbour. Some experts consider that there cannot be too many of them. This is probably a mistake. As before stated, such a light should act direct from its lantern when used for searching purposes; but when a hostile flotilla comes to close quarters, these expensive lanterns should be lowered under cover and an arrange-

Fig. 93.

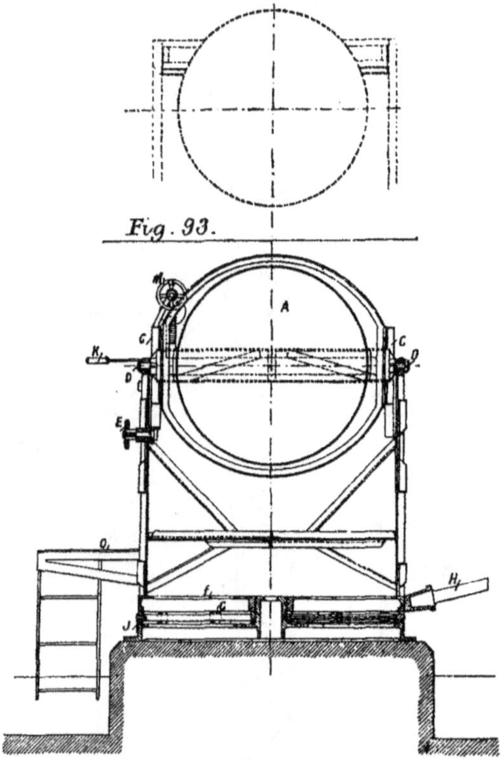

ment raised whereby the ray of light can be reflected in any desired direction by a cheap plane mirror that can be readily replaced when broken.

The following arrangement has been worked out by Messrs. Day, Summers, and Co., Southampton, from a design by the writer. The apparatus is placed in a pit behind a parapet. A masonry pedestal in the centre of the pit (see Figs. 92, 93, and 94) carries a pivot round which a turntable F revolves on rollers G. The pivot is secured to a ring on which G G roll, and round which is placed a brake strap for holding the turntable by the handle K whenever required. The turn

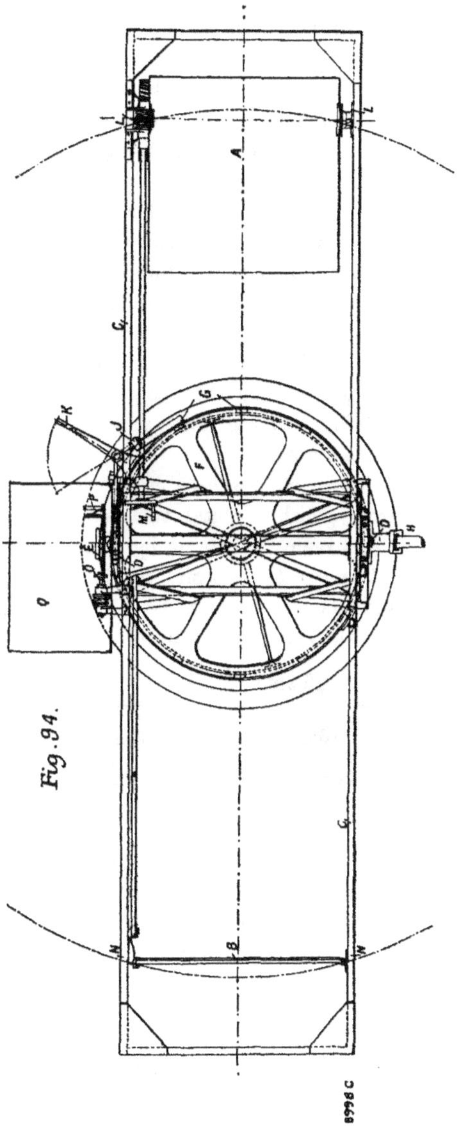

Fig. 94.

table can be revolved by a handspike H directly K is released. A carriage, fixed to the turntable, supports on the trunnions D two girders framed together at their ends and connected by a ring at the centre. This frame can be clamped in any position by the arc and screw E. An electric lantern A is supported at one end of the framed girders on trunnions L, and is capable of vertical adjustment by the handwheel M. A plane mirror B is supported on trunnions N at the other end of the frame and can be adjusted vertically by the handwheel O, or, if flashing signals are required, the mirror can be moved quickly in altitude by the handle P. A platform Q, bolted to the carriage, supports the operator.

When the apparatus is out of action the frame is brought into a horizontal position, and everything is then under the crest of the parapet. When it is to be used as a search light the lantern end of the frame is raised and the light works direct from its lantern over the crest of the parapet. When it is desired to lower the lantern and protect the light the carriage is turned round, the lantern lowered, and the mirror raised. The electric connections are not shown. There is no difficulty. The wires pass through the centre of the pedestal and the pivot and are carried direct to the lantern, thus avoiding any sliding connections outside the lantern.

There is a certain loss of light when a mirror is used. The quantity of light reflected from polished metal surfaces is greatest when the angle of incidence (between the ray of light and the normal to the surface) is small, but an exactly opposite result is obtained with non-metallic surfaces, such as water, glass, &c. The percentages of light reflected from various surfaces are as follows :

TABLE XXXIII.

Angle of incidence ...	75 deg.	60 deg.	45 deg.	30 deg. and under
Water	21 p.c.	6.5 p.c.	...	1.8 p.c.
Glass	30 p.c.	11.2 p.c.	4.5 p.c.	2.5 p.c.
Polished silver ...	nearly 90 p.c.			

Glass with a silvered back acts nearly as efficiently as polished silver, and the silver does not tarnish, the air being thoroughly excluded from it. Glass is therefore greatly preferable to metallic mirrors. At long ranges there is a loss of light due to dispersion caused by want of absolute parallelism between the two glass surfaces, and I have calculated that this loss reduces the efficiency of a glass mirror as per following Table :

TABLE XXXIV.—GLASS MIRROR SILVERED ON THE BACK.

Angle of incidence ...	75 deg.	60 deg.	45 deg.	30 deg. and under
Efficiency at short range...	88.9 ,,	89.7 ,,	90.5 ,,	90.9 deg.
,, ,, long ,, ...	59 ,,	77 ,,	84 ,,	86 ,,

These results were obtained thus :

Let a ray of light, power 100, fall on a glass mirror at an angle of incidence 45 deg. (see Fig. 95), 4.5 per cent. will then be reflected (see

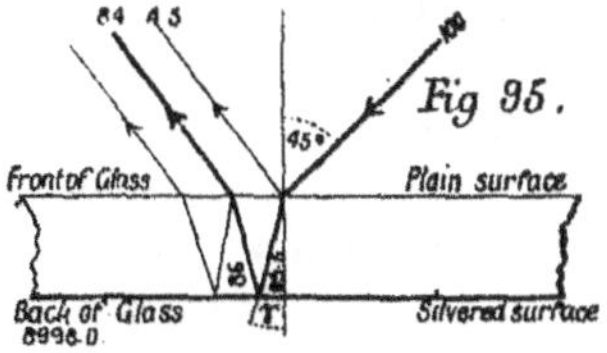

Fig 95.

Table) from the front surface, and 95.5 per cent. will penetrate the glass to the back surface in a direction depending upon the index of refraction (about 1.53 for plate glass), so that the

$$\text{Sin angle of incidence} \div \text{sin angle of refraction} = \text{index},$$

or in above example,

$$\text{Sin } 45 \text{ deg.} \div \sin a = 1.53,$$

consequently

$$a = 27 \text{ deg. } 30 \text{ min.}$$

From Table XXXIII. the amount of light reflected from the silvered surface will therefore be 90 per cent. of 95.5, say 86, and nearly 84 of this will go away from the top surface of the glass nearly parallel to the 4.5 ray. As, however, the glass surfaces can never be absolutely parallel, these two rays, as well as the smaller third ray, will gradually diverge, and the power of the reflected ray at 45 deg. will therefore never exceed 84 per cent. except at short ranges.

Similarly, it can be shown that the principal reflected ray for angles of incidence of 30 deg. and under, and 60 deg. and 75 deg., are 86 per cent., 77 per cent., and 59 per cent. respectively. These figures demonstrate the advantage of so using glass mirrors, that the angle of incidence is small. The arrangement already described does this, the mirror being so placed that the reflected ray when directed on the horizon from an emplacement near the water level makes an angle of 36 deg. with the principal axis of the frame carrying the mirror and lantern, the angle of incidence being only 18 deg. Not less than 86 per cent. of the light is therefore reflected (see Table) when the outer glass surface is clean and dry.

Means should be provided for rapidly replacing the plane mirror if it be shot away. An arrangement designed by Major M. T. Sale, C.M.G., R.E., in which a disc of silvered copper is stretched on a frame like a drum-head, has been found to answer admirably as a plane mirror. It remains effective after being hit by bullets or shrapnel. This was proved conclusively by recent experiments carried out at Okehampton and reported in the *Times*.

Passing to the lantern, the catoptric arrangement designed by Colonel Mangin, of the French Corps du Génie, is probably the best, although it is said that a good dioptric lamp has been known to beat it. But the focal distance of the carbons is so much shorter in most of the lanterns fitted with concentric prismatic rings of glass that it is most difficult to obtain and keep the light at true focal length. In the catoptric lanterns the focal distance can be much longer without greatly increasing the cost. Moreover, fairly good results can be obtained by cheaper arrangements than Colonel Mangin's, one of the best of these being the result of Captain Cardew's scientific labours at Chatham. In these less costly designs for a catoptric lantern, efficiency appears to depend upon the size of the curved reflector, and it is interesting to note that similar results were obtained in 1855, as the outcome of the investigations made by General Cator's committee, appointed by the Director-General of the Ordnance. After trying various forms of parabolic and other reflectors, it was then agreed that the best was a large spherical surface about 3½ ft. in diameter, and forming 50 deg. of a sphere 8 ft. in diameter. The long focal distance obtained from the use of such a spherical surface appeared to give better results than the theoretically perfect surface of a paraboloid where the focal distance is necessarily shorter; and this is especially true when the electric light is used, the carbon points occupying a certain dimension. The sine of the angle of dispersion is equal to the radius of the dimension of the source of light divided by the focal distance, and this consideration demonstrates the advantage of using a lantern (whether catopric or dioptric) with a long focal distance.

The local glare caused by a powerful electric light is so great that the observer should be located at some distance from it, and it may frequently occur that he should be stationed well to the front and afloat. The light should be under his control, if possible, and the simplest way to do this is to erect two single needle galvanometers on the carriage of the apparatus, and to connect them with the observer. A deflection to the right or left on one needle can indicate, elevate, or depress; and on the other needle traverse right or left. In this manner a search light on shore may be under the control of an officer in command of the defence flotilla, the steamer being anchored in a suitable position and an electric cable taken to it. One or two good search lights are probably sufficient.

In addition there should be one or two electric lights for providing a fixed ray across the waters in advance of the mine fields. These lights should not be displayed until the attack has been pushed to the waters in front, and until the defence flotilla has retired in rear of the ray or lane of illuminated water. The defence will then be enabled to open a heavy fire on any attacking boat that may venture across the lighted area.

When the attacking forces have so far developed their operations the defence would derive great assistance from the employment of powerful floating lights, and the Lucigen Light Company is endeavouring to perfect an apparatus of this nature. A "triplex Lucigen" has a power of 10,000 standard candles, and is stated to be capable of illuminating an area equal to a quarter of a square mile—say a circular area half a mile in diameter.

Probably the best way to use the Lucigens for illuminating harbour entrances will be to fit them on small steam pinnaces, each so built that the boat cannot easily be seen or struck, the hull being nearly submerged, and having a steel-faced turtle-backed deck. The air-compressing machinery required for the light could be driven by steam taken from the boiler for the propelling machinery, and the boat could easily be fitted with an air receiver, and with a tank to carry oil for replenishing the Lucigen as required.

Such a boat could be moored suitably, so that in the event of an attack, the powerful lights carried would be displayed, and the attack be thereby clearly seen by the gunners of the defence, both on shore and afloat.

Electric lights are most useful for search and discovery work, but they illuminate small areas intensely, leaving the rest in deep shadow. If they were supplemented by lights like the Lucigen, directly an attack developed, the whole of the water could be illuminated and the power of defence be greatly augmented. There should be no difficulty in screening the light of each Lucigen, so that it would be directed only in a desired direction, and the armed guard boats of the defence might take up positions in the dark area in rear of the lights as soon as an attack had been discovered, either by observers on shore or afloat. The attack would thus be delivered under great disadvantages, for a well-lighted area would have to be crossed, which would be swept by the fire of machine guns afloat, and of quick-firing cannon in the shore batteries.

Obstructions.—Just as the mines themselves form a grand obstruction to the passage of large vessels and thereby assist greatly in the defence of a maritime fortress, so smaller obstructions can be usefully employed to impede and perhaps prevent the passage of small craft whose aim may be to attack the mines.

Floating Nets can be used, and if small steamers get among them the propellers are often fouled, the boats become temporarily helpless, and may be destroyed by a well-directed fire from quick-firing or other guns.

Nets can, however, be readily passed if seen, small lengths of chain being thrown upon them from the boats, which effectually sinks them.

Booms, when well made, are probably the most effective passive obstruction against the passage of boats. They should invariably be formed of a double line of balks at such a distance apart that a boat which succeeds in jumping or passing through the front line, is brought up by the second line. The two lines should be connected frequently by cross-beams, the whole presenting the appearance of a large floating ladder. Wire ropes or chains should run along each line and be connected to the main anchors at the ends of the boom. Stream anchors should also be connected to the boom at intervals in order to keep it in position when the tidal or other currents flow across it, and during stormy weather. If such a boom be deficient in buoyancy, empty casks can be lashed to it at intervals. An attacking flotilla generally tries to destroy such an obstruction by means of small charges attached to it and then fired by electricity through a length of insulated wire. Some experts consider that such a boom should be moored on the waters immediately in front of the mined area, and that the fixed ray from an electric light should be directed upon it or upon the water immediately in front of it. It is perhaps better to place it, as well as the ray of light, some distance in front of the mines, whose position is not then demonstrated with such precision. The boom should be just behind the ray of light, but should boats attempt to fire charges on the boom the light should be thrown upon them, and every means taken to prevent the explosion of the charges. The boom should consequently be enfiladed by some quick-firing guns of the defence, and if a few circuit-closers be connected to the boom in such a way (to be described presently) that a signal is given on shore when a boat comes against the boom, the defenders will know when to open fire and how to direct it, even in darkness and when no object is seen. This idea is similar to "automatic artillery fire" advocated by the writer in 1883.* The approaches to a boom can also be sown with small mines, specially designed, and having a proper submersion for acting against small craft.

When there is but little rise and fall of tide, either electro-contact or automatic mines may be employed, care being taken when the latter are used that the defence flotilla keeps clear of them, leading lights on shore being employed for this purpose. But a very small tidal rise and fall causes such mines to become useless at high water, unless they are awash at low water, which is not permissible, as the mines, if electric, would be constantly signalling by wave action, and if automatic would destroy themselves in rough weather.

* See paper read at the Royal United Service Institution, March 14, 1884.

In tidal waters Major R. M. Ruck's system of rise and fall mines (already described on pages 70 and 73) may perhaps be applied to boat mines, but it would probably be better and certainly less costly to attach mines to the boom itself, and to arrange so that each shall be exploded when a boat comes into contact. This action can be secured in the following manner.

The boom may be formed of a series of rafts of timber, connected together by two wire ropes or chains, one along the front, the other along the rear. To the front of each raft a wire is stretched, one end of this wire being fixed to a special form of circuit-closer which is actuated by a pull on the wire, the other end being secured to a spring that keeps the wire in tension. This spring can be secured to the back of the spar raft.

A single cable from a distant firing battery leads to each circuit-closer in turn, thence to the mine which that circuit-closer explodes. The circuit-closer can be placed either on the front or the back of the raft; the pull-wire in the latter case being led through a pulley at each end of the front spar.

The explosion of the mine does not injure the raft or the circuit-closer, but it destroys the boat whose bow causes the pull on the wire by coming into contact with it. Fig. 96 shows the arrangement.

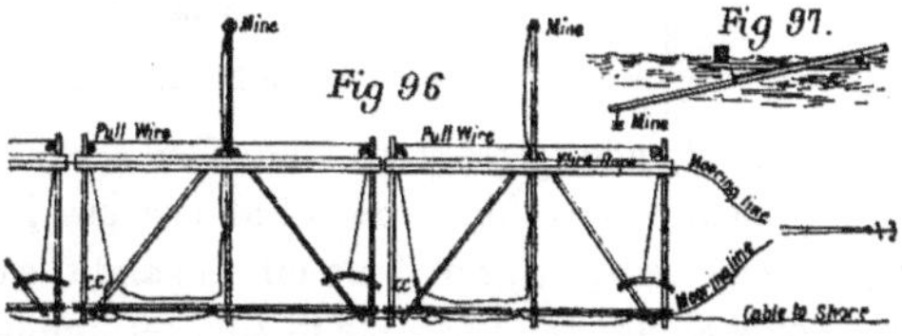

Fig 96

Fig 97.

The outrigger is provided with an iron eye about 8 ft. behind the mine, and a rope being previously secured to the centre of the front log and its end secured to the back log, this end is passed through the eye on outrigger, and the spar then thrust out; the inner ends of the spar and of rope are finally secured to the centre of the back log. The electrical joint between mine wire (which is stapled in a groove along the under side of the outrigger) and circuit-closer wire is then connected up. By these means another mine on its outrigger can readily be connected up in boom to replace one that has exploded.

Fig. 97 shows a sectional view of a raft and mine. The rafts would be connected up in line, on a beach, just below high-water mark, commencing work on a falling tide. The mine and its outrigger spar would be connected up separately when the boom is in place. The small insulated wire to circuit-closer and to mine would be fixed in grooves to the spars by small staples.

The rafts being connected together in boom, they are towed out as soon as the tide floats them, and taken to their position on the mine field, a large anchor having previously been laid and buoyed. One end of the boom is secured to the mooring line of this anchor. The boom is then stretched into its position, the wire ropes being made as taut as possible, and a second large anchor laid, and its mooring line secured to the far end of the boom. Stream anchors, when required, are now laid, and their mooring lines secured to the front and back of the boom, between the rafts.

The electric cable is then connected to the near end of boom, and is laid to shore. The raft mines on their spars are brought up by another boat, and are lashed in position, one on each raft. This operation can be performed at the same time that the electric cable is being laid.

By proceeding in this manner, most of the work is done on shore, and a great deal of it can be done permanently, the rafts being stacked at the depôt ready for use at any moment.

The principal cost of these arrangements is that of the spars and materials used in the boom. As before stated, a boom of some kind is essential to ward off boats, and small mines of some sort in front of the

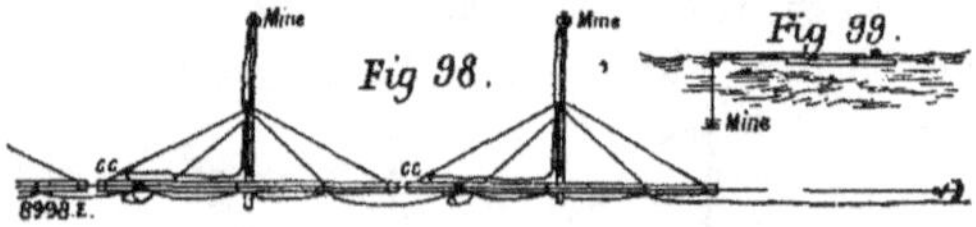

Fig 98. *Fig 99.*

boom or attached to it, add greatly to the efficiency of the defence. The above arrangement is therefore economical, because it obviates the use of sinkers, mooring lines, &c., for these mines; and the circuit-closers and mine cases being almost invulnerable in countermining, the defence against boat attack, produced by this design, is strong. It moreover provides against the difficulties engendered by the rise and fall of tide.

In the event of the number of spars available being limited and insufficient to form a boom, as described, the following alternative design may be used. It is more economical in material, but not so trustworthy. This arrangement is shown in Figs. 98 and 99. It consists of a log and a cross-spar projecting a few feet beyond the centre of the former, and secured there by means of two wire ties.

An outrigger spar is lashed to the cross-spar, and projects to the front a few feet more.

The circuit-closer and pull-wire form a projecting triangle to the end of the cross-spar, to which they are secured by a small pulley, not shown on the drawing.

Elasticity in the pull-wire can be obtained by securing the pulley to the cross-bar end, by means of a strong india-rubber strop, or by any other simple spring.

The mine is suspended from the end of the out-rigger spar.

The electrical arrangements and the mode of mooring are similar to those already described for the boom, composed of rectangular rafts.

The stream moorings would have their lines secured to the wire rope at the point of junction of two rafts, and the end moorings, and electric cable to those, would be arranged precisely as in the former example.

The circuit-closer is shown on Figs. 100 and 101. It consists of an iron tube provided with a movable end-piece, to which is secured one

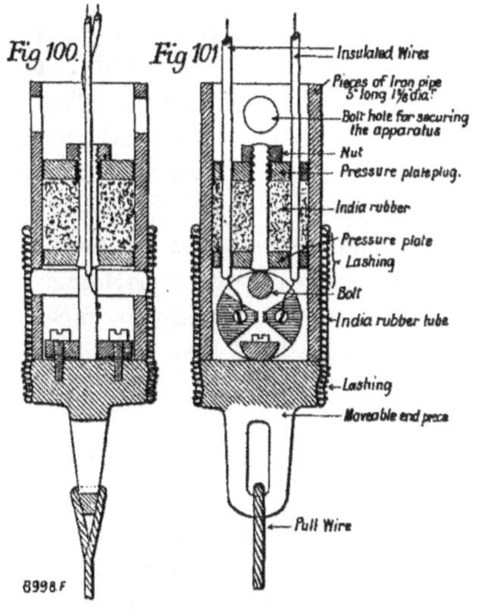

of my patent spring ring contact-makers. The other side of the ring is held by a bolt that crosses the iron tube, the ends of the bolt lying flush with the outside surface of the tube. The two electric wires are led through a pressure plug consisting of two iron discs, an india-rubber plug, and a screw bolt and nut for compressing same, the nut being turned by a box spanner from the tube end. By such an arrangement, the plug forms a good water-tight joint both for the tube end and for the wire entrances. Also, the plug rests solidly on the before-mentioned cross-bar, and cannot therefore be driven in on the contact spring by the pressure produced by the explosion of a countermine, or of a neighbouring mine.

The end-piece and part of the iron tube, are surrounded by an india-rubber tube lashed to each of them. This allows the end-piece to be pulled out for a short distance by a pull on the wire, and the elasticity of the india-rubber tube helps the elasticity of the spring ring to pull the end-piece back to its normal position, thus reopening the contact points, and insulating the branch wire from the electric cable if the mine be fired, or bringing everything back into the normal condition if the mine be not fired. The latter occurs when the firing battery has been purposely disconnected, in order that the system may be tested by bumping each raft in turn by a defence boat, a test battery and galvanometer only being in circuit on shore, during such an operation.

The circuit-closer is fixed to a spar or other object by means of a cross-bolt carried through a hole in the tube at the end furthest from the end-piece. In order to prevent the circuit-closer being unduly strained by the bumping tests, two side links (not shown on the drawing) are provided, which hinge on the long cross-bolt. These links connect with another cross-bolt engaging in the hole provided in the end-piece for the pull-wire, the hole being made of a sufficient length and size for this purpose (see Fig. 101).

The arrangement of end-piece, &c., is designed to form an efficient protection to the apparatus against countermining, but it is of no use to make the circuit-closer, raft, and pull-wire impervious to damage by countermines, unless the mine be so also. The inventor has, therefore, taken much trouble to design an arrangement for the charge which shall be safe against countermining.

The charge, about 20 lb. of wet slab gun-cotton, is firmly braced between two iron plates by bolts and nuts, and the priming charge of dry gun-cotton is placed in a short length of boiler tube, the end of which is closed by an insulated plug for the wire entrance. This plug has a circular rim between which and the tube end an external leather washer is pressed by two small studs secured to the top plate of the charge. The wet gun-cotton is cut away centrally so as to fit against the tube containing the priming charge, and a hole in the top plate coincides therewith. A 20 lb. charge will act effectively against boats to a distance of 10 ft., and the rafts can therefore be from 20 ft. to 25 ft. in length. The apparatus is manufactured by Messrs. Elliott Brothers, London.

Cribs of Timbers filled with stones, and other obstructions of the kind, can be used when it is desired to prevent boats passing over shallow waters.

Boat Defence.—But the best defence against boat attack is an active boat defence. Should this collapse or be non-existent, the best systems of passive obstruction must fall before an enterprising foe.

Smokeless Powder.—The employment of such powder (see page 209) is now being carefully tried by Lord Armstrong, who in a recent speech said:

"A new departure has been made in the manufacture of powder . . . . for our quick-firing guns. It is made by the Chilworth Company . . . . and notwithstanding that the charges have been reduced in weight by about one-third, we have obtained velocities of from 2300 ft. to 2400 ft. per second, as compared with 2000 ft. . . . . with other powders. This new powder, moreover, leaves no residue in the gun to interfere with the essential requirement of rapid loading, and the smoke has been so far reduced as to present little obstacle to the sighting of the guns in action.

"These are advantages which can scarcely be over-estimated."

CHAPTER XXI.

TORPEDOES.

As submarine mining officers not unfrequently have the charge and direction of torpedoes employed in harbour defence and actuated from shore, a treatise on submarine mining would not be complete without a chapter on this subject. The following information is chiefly derived from articles published in *Engineering*, 1887, 1888.

Torpedo Batteries.—The employment of torpedoes in batteries specially constructed for them has often been recommended for the defence of narrow channels, entrances to harbours, &c., and this method has been adopted by some nations in certain favourable situations.

The Whitehead Torpedo has been adopted as a naval arm by so many nations, and has received so much attention for a long period of years, that it has probably been brought to the maximum state of efficiency obtainable from it. It is so well known that a detailed description is unnecessary, but it will be as well to note some of its chief characteristics, and especially those which are defects inherent to the invention.

The case, formed of steel, or of a very strong alloy,* is from 14 in. to 16 in. in diameter, has a length of about 10 diameters, is circular in cross-section, and is pointed at each end. It weighs about 600 lb., and carries a charge of from 40 lb. to 70 lb. of high explosive. It is propelled by two screws, one abaft the other, worked in opposite directions and driven by a self-contained engine and a reservoir of highly compressed air possessing a potential energy of about $\frac{1}{4}$ million foot-pounds. A regulating valve causes the engine to be driven at any desired speed. This valve can be so adjusted that the mean speed of the torpedo may be 25 knots for a range of 200 yards, or 22 knots for a range of 600 yards, or intermediate speeds for intermediate ranges. The charge is carried in that portion of the case near the head, which is fitted with an apparatus that causes the explosion when the torpedo strikes the side of a vessel, point first. The immersion of the torpedo is

* The Schwartzkopf Whitehead is essentially the same as the Finnie Whitehead, but is made of phosphor bronze instead of steel.

regulated by horizontal rudders at the tail, and these are actuated by compressed air, governed by a valve, itself controlled by the hydrostatic pressure due to the immersion, and by an attached pendulum weighing from 30 lb. to 40 lb. The method of regulating lateral direction is by vertical fins permanently adjusted in accordance with experiments made with each torpedo. The Whitehead is ejected by compressed air or by the explosion of a small charge of gunpowder, the directing tubes, carriages, or other apparatus varying according to the conditions of each situation, such as under-water or over-water discharge, and front or broadside discharge. These torpedoes can also be discharged by gravity, like the ball of a falling pendulum, release being effected at or near to the lowest point of the fall, or by running down an inclined plane curved somewhat in the form of a parabola, viz., steep at first, then gradually becoming nearly horizontal as the torpedo reaches the water surface. This latter method appears to be specially favourable for employment in shore batteries. It was tried at sea, but discarded owing to the irregularities caused by the pitching motion of vessels in a seaway. The defects of the Whitehead torpedo in the order of their importance are:

1. *Inefficiency* due to the small charge carried, which is now insufficient to destroy the hulls of vessels like modern ironclads that are divided into numerous water-tight compartments.
2. *Uncertainty as to Accuracy.*—For, although a vessel can generally be hit up to a range of 300 yards, this cannot be depended upon the course of a Whitehead occasionally being very erratic, especially with over-water discharge from the broadside of a vessel at speed. Moreover, during handling and discharge, the fins, and rudders, and other gear projecting from the body of the torpedo, are liable to derangement. Inaccuracy as to submersion is also encountered, due to imperfections in the design or manufacture of the automatic controlling gear.
3. *Expense.*—The manufacturing cost of one Whitehead being over 500*l.*, to which must be added the share of price first paid for the patent, and the cost of the discharging appliances.
4. *Intricacy.*—The torpedo containing a quantity of highly finished and complicated machinery.
5. *Difficulties in Manipulation.*—Great intelligence on the part of the personnel combined with a long and careful training, being essential.
6. *Difficulties in Maintenance.*—Constant attention and care being required to keep the torpedoes and their impulse arrangements clean and efficient.

7. *Loss of Control after Discharge*, which, combined with the uncertainty as to accuracy already mentioned, increases the difficulties attending the employment of these torpedoes in fleet actions.
8. *Motive Power Dangerous*, the highly compressed air having sometimes burst the torpedo. Hostile shot would increase this danger.
9. *Space Occupied*, especially when that of the appurtenances are taken into consideration.

Not only are the above defects recognised by many critics whose opinions are not to be despised, but the torpedo boats specially built to carry the Whitehead are now regarded with much less favour than formerly, owing to the physical impossibility that human beings can live on board when the boats are required to keep the sea for any length of time. Indeed, it appears that all Whitehead torpedo boats that are too large to be hoisted on board a man-of-war, and yet too small themselves to keep the sea, must be relegated to harbour or river defence.

Contradictory as it may seem, defect No. 3—the great cost connected with the Whitehead—has been the means of perpetuating its employment. After spending vast sums of public money on any engine of war, those responsible are loth to acknowledge its defects, and prefer to spend more in attempting to perfect the invention. As regards the Whitehead we are in the same boat with most of our neighbours, and we were almost compelled to act as we have done, but it is high time that other inventions should be carefully examined and compared with it.

The Howell Torpedo.—This, the invention of Captain Howell, United States Navy, is similar to the Whitehead both in outward appearance and in general design, Fig. 102. The charge is carried in the forward cone, the motor in the centre of the body, twin screws and horizontal and vertical directing rudders aft. The most important novelty is the motor, which is simply a ponderous steel gyroscope on a horizontal axis across the centre of the torpedo. See Figs. 103 and 104.

A torpedo 8 ft. long and 13.3 in. in diameter carries 70 lb. of explosive, and a flywheel of 110 lb., the whole torpedo weighing only 325 lb. A Howell torpedo, as heavy as the Whitehead, and 14 ft. 6 in. long, will carry over 200 lb. of explosive.

The flywheel is spun up to a speed of 10,000 revolutions per minute, over half a million of foot-pounds being then stored in the motor. A Barker's mill is generally employed to perform this work, the flywheel axle being grasped externally by a clutch on the driving shaft of the Barker's mill, and being disengaged when desired.

The shafts of the twin screws are connected to the flywheel axle by mitre wheels, and it is stated that an 8-ft. torpedo can be driven by a 110 lb. flywheel at a speed of 24 knots for 600 yards.

"The fundamental principle upon which the steering of the torpedo

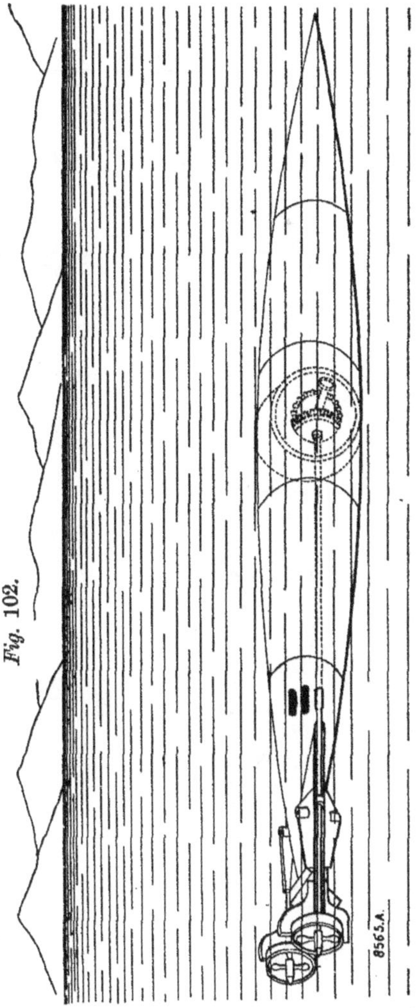

Fig. 102.

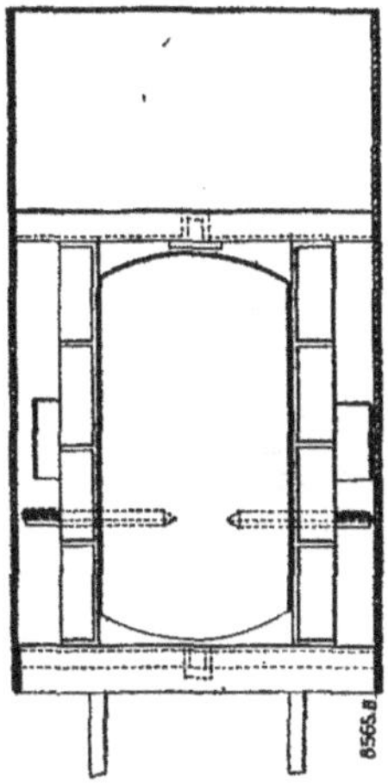

Fig. 104.

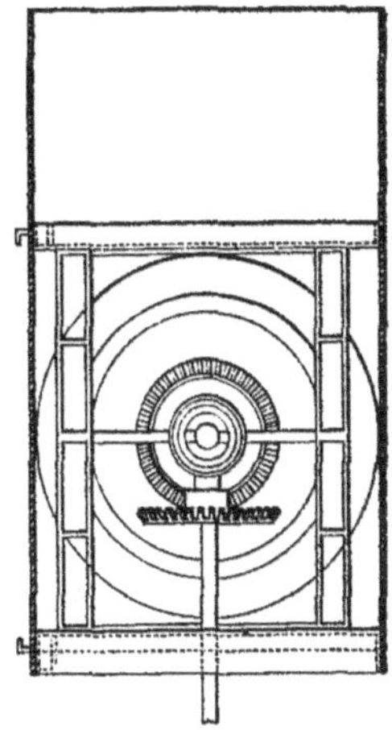

Fig. 103.

is based is that if a revolving flywheel be acted upon by any force which tends to turn it about any axis not parallel to its own, there will be a resultant motion about an axis perpendicular to the plane of those two. This offsets and opposes lateral deflection of the torpedo, and compels it to travel in the course in which it was originally pointed or launched. The axis of the flywheel being horizontal, any extraneous force tending to deflect it laterally will cause the torpedo to roll, which rolling can be conveniently employed to bring into action steering mechanism arranged to apply automatically an opposite deflecting or deviating force which will restore the *status quo.*

Fig .105.

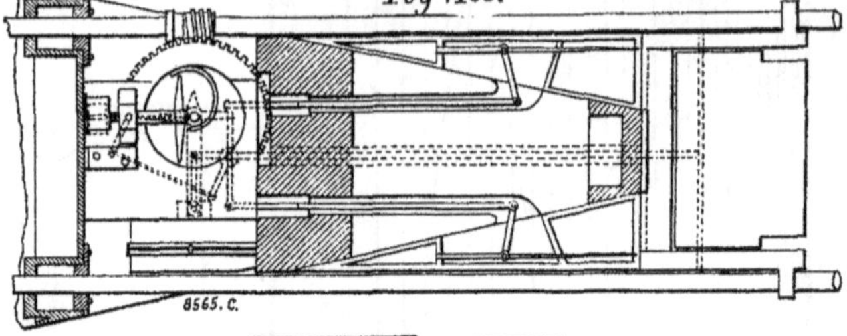

Fig 106.

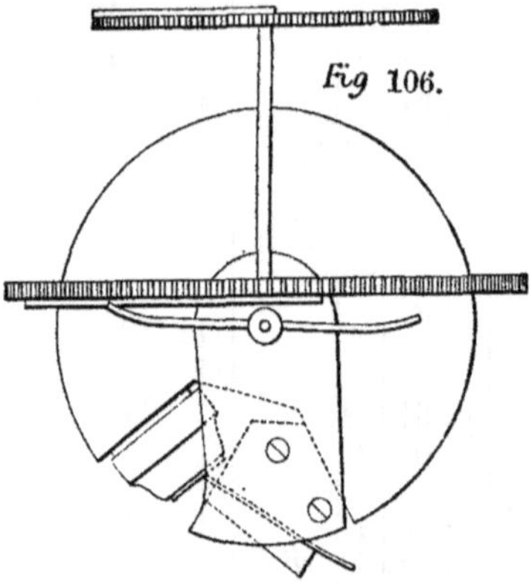

Fig 107.

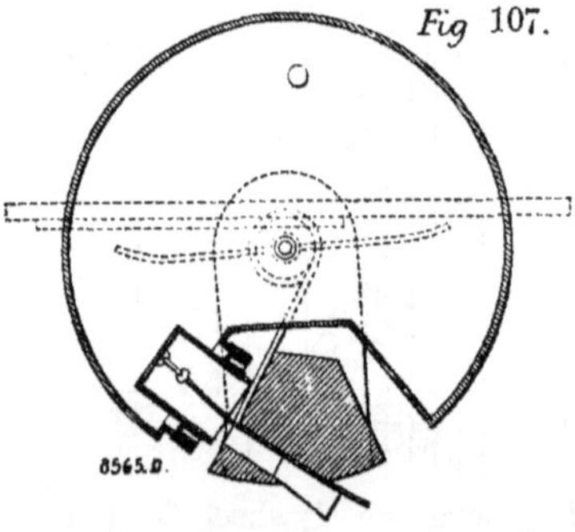

" The steering mechanism, Fig. 105, consists of one or more vertical rudders and rudder-operating devices, so arranged that when the torpedo rolls to starboard, the helm automatically will be put to starboard and *vice versâ.* As the horizontal axis of rotation of the flywheel is transverse to the longitudinal axis of the torpedo, it is necessary to provide a diving rudder to keep the torpedo during its run at a given depth. This rudder is operated automatically by mechanism, Figs. 106 and 107, the action of which is controlled by a combined pendulum and regulator, the latter being governed by the pressure of the water which varies with the immersion of the torpedo. The office of the regulator is to cause the torpedo to sink and maintain itself at the required depth; that of the pendulum to prevent the torpedo from diving or rising too abruptly.

"At the after end of the torpedo, surrounding the propellers, are tubes which, by reason of the mass and velocity of water flowing through them, serve to stiffen the path against irregular movements in the vertical plane.

"The discharging gear used up to the present time consists of a frame, or derrick, extending from the ship's side under which the torpedo is hung by clutches and studs on its shell. The frame is either pivotted on the rail or fitted to slide in and out on a stationary beam; in either case the torpedo can be slung from the deck, then rigged out and operated with rods or lanyards, steam being turned on the Barker's mill, and the wheel spun up; one lanyard acting on a trigger, disengages the clutch connecting the two; the other lanyard, acting also on a trigger arrangement, disengages the torpedo from the clutches. To give it an impulse in the direction in which it is launched the torpedo is also grasped abreast the centre of gravity by a downward switching clutch, pivotted outboard on the frame beyond; on being detached from the derrick, it is swung outboard in the arc of a circle and detached automatically by a check and trigger on reaching the vertical below the pivot. This gives it an impulse without changing the angle of its horizontal axis with the surface of the water. The supporting frame is free to swing below an axis parallel to the fore and aft line of the torpedo, so the axis of the flywheel is also kept horizontal.

"An improved apparatus, however, comprises a tubular shield protected by armour, in which the torpedo will be placed. At the inner end are two cylinders whose piston-rods reach forward and press against studs on the middle body. The tube and support revolve about a centre to allow lateral strain, the power for revolving the flywheel being conducted through this centre. Steam from the Barker's mill exhausts back into the condenser, thus stopping the humming sound, to which great objection had been justly raised. By one action of a lever the power is shut off and the torpedo ejected."—*Engineering*, January 20, 1888.

The objection has been raised that this torpedo "does not lie in a state of constant readiness, but has to be spun up" before it is ready to launch, but it must be noted that when the wheel has been spun up, very little power will keep it going, and therefore the torpedo can be kept in the state of "ready" from the commencement of an action until its termination, unless, in the mean time, it be discharged.

Remembering the defects of the Whitehead torpedo which have been enumerated, it will be found that most of them have been overcome in the Howell torpedo.

Thus :

1. The efficiency due to small charge carried has been met.

2. Also the uncertainty as to accuracy.

3. Also the great expense, for the Howell torpedo and its appurtenances are cheaper to manufacture.

4. Also, *simplicity of detail* is substituted for that intricacy and delicacy of detail which in the Whitehead enlists our astonishment and admiration.

5. As regards *manipulation*, comparative trials are required, the advocates of the new arm being confident of the result.

6. The *maintenance* of the simpler apparatus must be less troublesome and costly.

7. The new arm is evidently under *better self-control* after discharge.

8. The danger due to the existence under fire of a chamber full of highly compressed air is absent.

9. And finally, the space occupied is less than with the Whitehead.

In short, it would appear that the Howell is superior on nearly all points, and, on account of its humming sound, is inferior only as an arm for a sneak boat, or for a vessel attempting to run a blockade.

The torpedo has been officially tried in the United States, and the Naval Board detailed to carry out these experiments has, it is understood, reported very favourably on the invention.

If used for harbour defence these torpedoes might be placed in shore batteries, and their simple fittings and accessories would not be difficult to keep in order. But it would generally be preferable to mount them on some floating body and moor it under the shelter of the land or a fort in a convenient place for aiding the defence. By these means, a foe would be kept in ignorance of the position from which his vessels might be torpedoed should they attempt to force a passage.

CHAPTER XXII.

Controllable Torpedoes.

The next class of torpedo to be considered is that which is controlled after discharge and is directed to its object from the base whence it is launched.

For many years it has been seen that a successful weapon of this nature would be useful in certain situations for harbour or river defence; and its more sanguine admirers believed in its becoming an important naval arm, but at present the best-known forms of controllable torpedo find no favour in our own or other navies.

The Lay, the Ericcson, the Berdan, the Sims-Edison, the Nordenfelt, the Patrick, the Lay-Patrick, and the Brennan are those which have received the most attention.

The Berdan is propelled by the gas from burning rocket composition; the Lay and Patrick by compressed carbonic acid gas; the Ericcson by compressed air; the Sims-Edison by electricity located at the base; the Nordenfelt by electricity carried in the torpedo. Nearly all are controlled by electricity acting on valves or on electric motors.

The Brennan Torpedo, however, is propelled and controlled without gas, air, or electricity, and it carries but little machinery, for the engine that propels it is stationed at the base of operations.

"The mode of propulsion is effected by the rapid unwinding of two wires from two drums or reels carried in the interior of the torpedo, and connected respectively to the two propeller shafts, thereby causing the two propellers to revolve at a high rate of speed, and consequently forcing the torpedo through the water. The unwinding of these two wires is effected by means of a powerful winding engine placed at the starting point on shore. Considerable interest has been evinced in this invention since its first appearance, because of the apparent paradox involved in its mode of propulsion, in that the harder this torpedo is pulled back the faster it will go ahead; but on consideration it will be seen that by hauling in the wires at a certain rate, a corresponding rate of revolution is imparted to the drums which are fixed to the propeller shafts in the torpedo, and so to the two propellers, which are thereby capable of developing a certain horse-power, and if this horse-power be sufficient to overcome the retarding strain on the wires, and to leave a margin of thrust, then the torpedo must be propelled through the water; and the only limit to the speed of the torpedo is apparently the strength of the wires."

The principle involved is similar to one embodied in a gun-rammer which Lieutenant (now Major) T. English, R.E., invented many years ago, and brought to the notice of the Ordnance Select Committee. It consisted of a small carriage engaging the bore of a gun by rollers driven by a chain so that the carriage or rammer was driven down the bore on the chain being pulled in the contrary direction.

Some doubts have arisen as to the accuracy with which the Brennan torpedo can be steered, but the personal equation enters largely into this matter, and those who know it best and are well able to judge of its capabilities are satisfied that it is sufficiently accurate.

Although the greatest care has been taken to guard the secrets of its construction, a very clever guess at its main features was published in *Engineering*, June and July, 1887, whence the following description and the above quotation are abstracted by permission of the editor.

Fig. 110

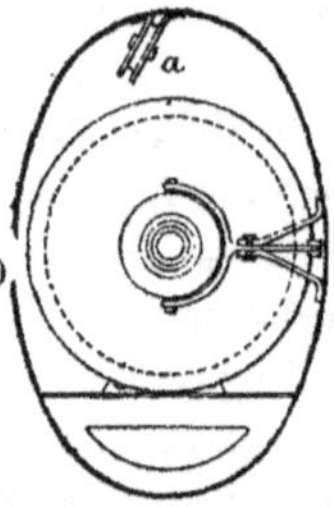

" Fig. 108 shows a section of the torpedo; Fig. 109 is a plan of the torpedo; Fig. 110 is a vertical section looking aft through X Y; Fig. 111 is a general view of the winding engine; and Fig. 112 represents the mode of using the torpedo.

" The dimensions of the present Brennan torpedo are 25 ft. by 3 ft. by $2\frac{1}{2}$ ft.; weight, fully equipped, 25 cwt.; speed, about 20 miles per hour; range, from $1\frac{1}{2}$ to 2 miles.

" I. *Mode of Propulsion.*—In Fig. 108, A and B show the two drums, or reels, on which is wound the wire by the unwinding of which the torpedo is caused to travel through the water; the fore drum A is attached direct to the inner solid propeller shaft S, and the after drum B is fast on to the outer hollow steel propeller shaft S^1; these two drums, by the unwinding of the wire $w\ w^1$, are revolved in the same direc-

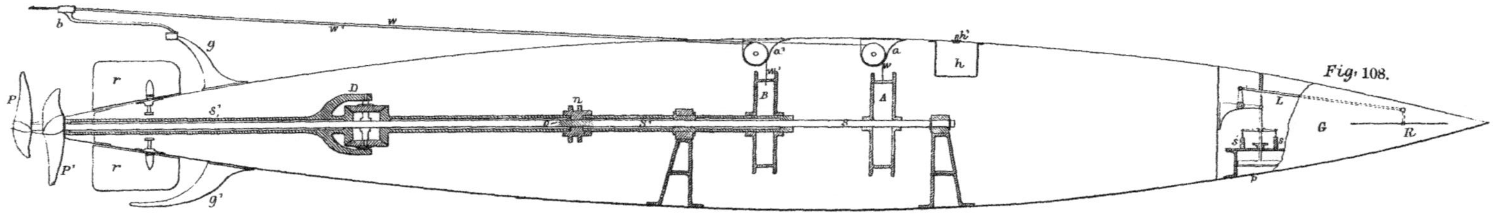

Fig, 108.

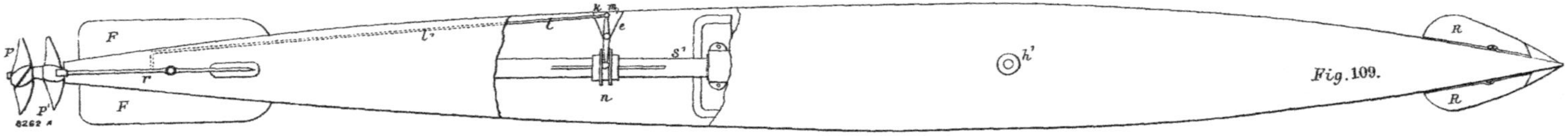

Fig. 109.

To face page 234.

tion, and their respective propeller shafts also, up to the point D; where, by a combination of bevel wheels (precisely similar to the arrangement adopted in the Whitehead, see Fig. 113), the outer hollow shaft S^1 has its motion reversed for the purpose of revolving the two three-bladed propellers P P^1 in opposite directions. At

Fig. 111.

first sight this appears a most unnecessary complication, if it be only required to effect the revolution of the two propellers in opposite directions, for this work could be more simply performed by taking the wires off the two drums, A and B, in opposite ways; but for the purpose of steering, the two propeller shafts should revolve in the same direction, while to enable the torpedo to maintain as straight a course as possible without utilising its rudder, the two propellers should be revolved in opposite directions, as was found so necessary in the case of the Whitehead. The two wires w w^1, are led from their respective drums

Fig. 112.

over the two sheaves a a^1, respectively, through the top of the torpedo by a hole made just large enough to take the wires, but without any gland. The wires then pass through a brass eye in the fair lead b swivelled to the guard g. The drums A and B are removed from the torpedo by withdrawing the inner propeller shaft S and taking out the fore drum A through a manhole in the side of the torpedo; the drum B being withdrawn from its propeller shaft and removed in the same way.

"II. *The Method of Steering.*—On the solid propeller shaft S is cut a screw thread, and immediately opposite, in the hollow shaft S[1], a slot is cut longitudinally. A nut or collar *n* with an internal thread fits in this slot, and on the hollow shaft. This collar *n* is grooved on the outside, and in this groove work two studs on the end of a forked lever *l*, Fig. 110; this forked lever *l* is carried by a bracket *m* on the side of the torpedo, and is connected at K, its other end, to a second lever *l*[1], which is in turn connected to the quadrant of the rudder shaft *r*. From this it will be seen that any movement given in a longitudinal direction to that part of the forked lever *l* which fits in the groove of the nut *n*, must transmit to the rudder *r* a movement to one side or the other. This longitudinal movement of the forked arm of the lever *l* is effected in the following manner. So long as the speeds of the two propeller shafts S S[1] (which up to the point D revolve in the same direction) be

Fig 113

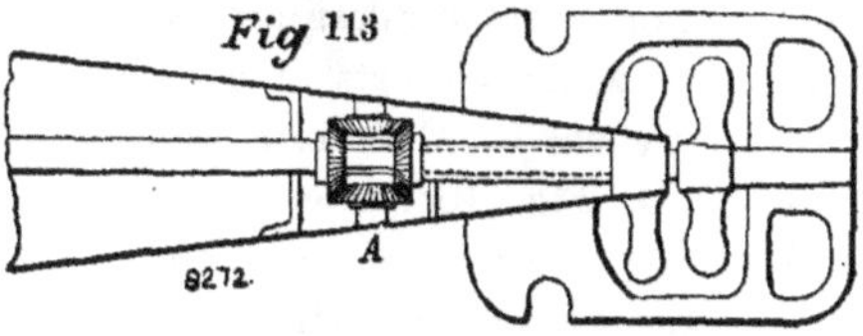

equal, the nut *n* with its internal thread, and the thread on the outside of the solid shaft at *o*, which engage, will revolve round together without any motion of the nut *n* along the shaft S[1]; but the instant a difference of speed is imparted to the two shafts S S[1], then the nut *n* will be screwed along the shaft S[1] either forward or aft, depending upon whether the thread is a right or left-handed one, and upon which of the two shafts is increased or decreased in speed as compared with the other one. Thus port or starboard helm can at any moment be given to the torpedo during its run at the will of the officer directing it, by altering the speed of one of the shore drums.

"III. *For Observing the Course.*—Several means have been tried to enable the operator at any moment to know the course the torpedo may take when running below the surface, among which may be mentioned a float attached by a line to the back of the torpedo, and a hollow mast with signal flag, but neither of these methods have proved very satisfactory, the former proving a very erratic indicator, and the latter taking too much away from the speed of the torpedo. It has now been found necessary to trust entirely to the use of phosphorus, or Holmes' light mixture, both of which when brought into contact with water emit flame and smoke in the track of the torpedo, the former being utilised for night, and the latter for day, runs. In Fig. 108, *h* shows the case in which this phosphorus, or Holmes' composition, is placed, and which is in connection with the water during a run by means of the hole h^1, placed immediately above the case *h*. The Brennan may be steered from 30 deg. to 40 deg. to port or starboard, but it cannot be turned round.

"IV. *Maintenance of Depth.*—To steady the torpedo in a submerged run, the two horizontal steel fins F F, Fig. 109, are provided. R R, Fig. 109, are two horizontal bow rudders, which by means of certain automatic arrangements are deflected up or down according as the torpedo reaches below, or rises above the depth it is set to run at. These bow rudders effect the same object in the Brennan that the stern rudders do in the celebrated Whitehead, and the only difference between the Whitehead and Brennan system of effecting the upward or downward motion is, that in the former there is an intermediate compressed air engine which actuates the stern horizontal rudders, but in the latter the bow horizontal rudders are directly actuated by the automatic arrangement, which consists of a balance weight or pendulum, and a hydrostatic valve. One form of this valve is shown in the bow compartment G, Fig. 108, where also is placed the balance weight. P is a piston exposed to the water on its lower face, and *s s* are two springs, the tension of which latter can be set to equalise the pressure of water on the lower face of the piston P for any particular depth at which it may be desired to run the torpedo. The movement of this piston up or down, corresponding to an increase or decrease of submersion, is transferred by means of a system of levers L to the rudders R R, and causes them to be deflected up or down, thus bringing the torpedo back to its normal depth. The Brennan is arranged to be run either on the surface, or from 8 ft. to 10 ft. below it.

"*Miscellaneous.*—In the fore compartment G, besides the automatic arrangements just described, is placed an ordinary self-registering instrument for recording the course of the torpedo on its run, as regards its depth below the surface at certain increments of time. Some of the records taken have registered great variations in the depth, very much more so than has ever been similarly registered by the Whitehead during the course of a run. This is only to be expected, as the longer and heavier Brennan would require more time to recover its proper depth and pass over more ground in doing so when once displaced. In actual warfare the charge of 200 lb. of gun-cotton would be placed in this fore compartment. G G^1 are two steel fixed guards to prevent the vertical stern rudders and the propellers from being fouled during a run of the torpedo by hawsers, chain cables, nets, &c.

" V. *The Wire.*—The steel wire principally used for the propulsion of the Brennan torpedo, is No. 18 W.G., breaking strain 6 cwt. to 7 cwt., weight per mile 33 lb. For any length of run three times the amount of wire is required to be wound on each drum; thus for a two-mile run, six miles of wire for each drum is needed, or twelve miles in all, equal to a weight of 392 lb., or 196 lb. of wire per drum.

" VI. *Operation of Winding.*—The various operations for winding the wires on the drums of the Brennan torpedo are as follows:

" 1. The wire is first wound off the reel on which it is supplied by the makers on to a split drum, *i.e.*, a reel or drum constructed to allow of the barrel being removed after the drum has been filled with the wire.

" 2. This coil of wire thus formed, having a hollow core, is placed in a tank of lime water, and carefully rinsed. If it be not required for immediate use it is left in this tank for several hours, so as to maintain the wire in a good state of preservation.

" 3. The coil of wire is removed from the lime-water tank and the wire wound on to a wooden swift—that is, a reel in the form of a cone placed on its base.

" 4. From this wooden swift the wire is wound on an ordinary drum, in a state of tension, preparatory to its final winding.

" 5. Lastly, it is wound off this ordinary reel to the torpedo drum. In this operation every care must be taken to insure the turns lying close together, so that the riding turns may not jam between any two of the underneath turns; and to lessen the chance of such a mishap, melted parafin wax is poured into every opening.

"On starting the winding of the wire in this final operation, the leading end is passed through a hole in one of the end plates, and fastened there; but on the winding being completed this end is cut free, with a view to prevent the wire bringing up suddenly on the whole length of it being run out. In such a case the wire is pulled out of the torpedo altogether. If for any reason the torpedo is stopped before the whole length of the wires has been run out, the wires must then be cut, and the four parts returned to the maker for rejointing up.

" VII. *The Winding Engine* (Fig. 111).—The drums, 3 ft. in diameter, are driven by a pair of direct-acting high-pressure engines, running at a great speed. Each cylinder is cast with the column under it, the latter being very strong and of such a form as to inclose the main working parts of the engine, and to prevent the wires from becoming entangled with any part of the engines in the event of the wire breaking. The steam is admitted by means of a valve common to both engines, and a governor is provided. The drums, running loose on the shafting, are connected by a 'jack-in-the-box' arrangement, by which their respective speeds can be regulated by means of a foot-brake without altering the speed of the engines. This 'jack-in-the-box' is arranged as follows: Cast solid on or bolted to each drum is a mitre wheel, and connecting these two mitre wheels are two smaller ones, revolving on their own centres, fixed on a carrier which is keyed to the main shaft. As soon as the main engine starts, the two small mitre wheels, which are in one with the shaft, are revolved with it, and carry round with them the two larger mitre wheels and consequently the drums. Hence it will be seen that on the brake being applied to one of the drums, the small mitre wheels will revolve round their own centres, the effect of which is to increase the speed of the other drum. As the speed of the one decreases so the other increases. The columns are braced together under the cylinders by another casting, and the whole stands upon a cast-iron sole-plate, thus making a very rigid formation. The engine is capable of working up to 100 indicated horse-power.

"VIII. *The Operation of Running.*—The torpedo is placed on a launching carriage, constructed in such a manner that the torpedo is automatically set free, and launched or pitched clear of it into the water, on the carriage reaching the desired position on the line of rails, laid on an incline to the water's edge. The hydrostatic valve is then set so that the horizontal bow rudders are correctly regulated, deflection upwards, for the particular depth and speed it is intended to run the torpedo at, the speed being governed by the number of revolutions given to the winding-in drums. The shore ends of the two wires are taken from the torpedo, secured to the winding drums, and one or two turns of the wires wound on. The carriage with the torpedo is then run down the incline, and the latter launched automatically into the water, the winding engine being at the same instant started. The torpedo then runs the required course and is guided by the movement of the stern vertical rudders to port or starboard. Its course is indicated to the observer on shore by the smoke in day runs, and in night runs by the light emitted from the composition in the case *h* (see Fig. 108).

"In the sketch Fig. 112, the winding engine is shown in a subterranean gallery in the fort F, with the line of rails laid to a point well beyond low-water mark. E is the engine, T a torpedo on its carriage ready for launching, T^1 a torpedo running its course towards the enemy's ship attempting to pass this earthwork.

"This completes the description of the Brennan torpedo and its *modus operandi*, and it will be evident to every one versed in torpedo matters that much of the success of this torpedo as regards three most important points, viz., "speed," "maintenance of depth," and "straightness of run," is by no means due to any special features in the original invention, but rather to the fact of its having been perfected at Chatham under the fostering care of scientific officers who are intimately acquainted with the details of the Whitehead torpedo with all its wonderful and clever mechanism.

"1. *Speed.*—The high speed of the Chatham-Brennan is due not alone to the clever mode of its propulsion, but in a great measure to the form and shape of its hull, and to the perfection of the winding engine; the former has been proved by Mr. Froude to be the most suitable form for obtaining high speeds for totally submerged vessels (it would be adopted for the Whitehead were it not for the necessity of a cylindrical compressed air chamber); the latter (the engine) has been specially designed at Chatham, and constructed by Messrs. Yarrow and Co.

"2. *Straightness of Run.*—Two screws revolving in opposite directions by which a wholly submerged vessel is more capable of maintaining by itself a straight course belongs to the Whitehead invention, as also does the system of gearing (shown at D, Fig. 108), by which this effect is obtained.

"3. *Maintenance of Depth.*—Horizontal bow rudders worked automatically by a hydrostatic valve and balance weight, together with fixed horizontal after fins, by the combination of which a wholly submerged vessel may be maintained during its run at a fairly constant depth, is also an adaptation from the Whitehead invention.

"4. *The Secret.*—Evidently the great and only secret in connection with the Brennan was the fact of its having been patented; a secret which has apparently been very well kept."—*Engineering*, 1887.

5. *Cheapness of Construction* is claimed by the advocates of the Brennan as one of its advantages, but a 100 I.H.P. steam engine is a costly item, and it can only operate one torpedo at one time, after which a period of preparation must elapse before a second torpedo can be discharged. Moreover the bomb-proof engine room, and the covered gallery to the water's edge, are costly; and if the torpedo itself be cheap to manufacture the right to do so was certainly dear to buy. On the whole, the advocates of this torpedo are scarcely justified in posing as economists.

6. *Range of Effective Action.*—Its range is said to be about 3000 yards, but it must be impossible to insure a strike at this distance except under the most favourable circumstances of weather, light, &c.; and then only if the operator be situated at a considerable elevation

above the water level. The engine should, however, be near to the launching way, and many difficulties are encountered when the operator is removed to a distance. The range of 3000 yards is often vaunted as a grand affair, whereas in reality the greatest objection to the Brennan is its limited sphere of action caused by the necessity of working it from a fixed point.

7. *The Wires* are easily broken by a kink or sudden jerk, but the history of the torpedo certainly proves that its success is principally due to the clever manner in which the wires have been pulled.

CHAPTER XXIII.

Torpedo Artillery.

The idea of projecting torpedoes to a distance by means of artillery is old, and a proposal to employ mortars for this purpose was brought forward officially in this country some years ago, but experiments were not recommended, high angle fire from mortars being considered too inaccurate for the purpose.

The correctness of this view may be open to doubt, modern *rifled* mortars giving very accurate results. The other great difficulties of the problem probably afforded the real reason for refusing to carry out any experiments in this direction.

The subject has, however, received the attention it deserves in America. Mr. Mefford, a schoolmaster of Ohio, was the pioneer. In 1883 he designed and constructed an air gun for throwing dynamite charged shells. It was 28 ft. long, 2 in. bore, $\frac{1}{4}$ in. thick, with an air reservoir of 12 cubic feet capacity, carrying 500 lb. pressure. Lieutenant (now Captain) E. L. Zalinski, United States Artillery, took up the invention, and has worked at it ever since with great skill and persistency. In this he has been backed by a strong financial company, and has been assisted by several high State officials, including the Secretary of the Navy.

An arm of great power and accuracy has been developed, the result being a complete vindication of the theories held by the advocates of air propulsion for torpedo artillery. The ballistic properties of the air gun are not sufficiently appreciated by many. In the first place, owing to the size of the air reservoir, the pressure exerted on the base of the projectile is nearly constant throughout the entire length of the bore; very different from the powder gun, wherein the pressure falls rapidly from the breech to the muzzle. Thus, in a powder gun of 16 calibres, the initial pressure being, say, 40,000 lb. per square inch, the mean available pressure may be under 12,000 lb. But in an air gun the length would be, say, 120 calibres, or nearly 7.5 times 16, and if 2000 lb. were used in a reservoir having ten times the capacity of the bore, the mean pressure would be 1900 lb., which, multiplied by 7.5, would be equivalent to 14,250 lb. acting through the *same length* as the powder gun.

In the second place, owing to the absolute certainty with which the air pressure can be regulated to any desired pressure, and to the uncertainty as to the pressure produced by the explosion of a gunpowder charge, the former is evidently much better adapted than the latter for discharging projectiles filled with detonating compounds. It is well known that occasionally, if rarely, exceedingly high and abnormal pressures have been recorded in experiments with heavy ordnance, no satisfactory explanation having ever been propounded except that of so-called "wave action." The possibility of such action makes a powder gun unsuitable for torpedo artillery. Again, the accurate manner in which range can be altered with the air gun by altering pressure instead of elevation is an important factor in its favour.

On the whole, therefore, Captain Zalinski may be considered to act wisely in adhering to air propulsion as the best method of solving the difficult problem on which he is engaged; and this, although the air gun possesses a great disadvantage in its length, and the space it necessarily occupies wherever its emplacement may be located.

Experience with the 2-in. gun already mentioned indicated that :—

"1. The valve should be automatic in its action as to opening and closing, and should permit the escape of a uniform volume of air between the two events.

"2. The length of gun should be as great as can be readily manipulated.

"3. The pressure should be at least 1000 lb.

"4. The gun should be easily trained."

A 4-in. gun was next manufactured, and many interesting experiments made whereby the fuzes and ammunition were improved.

In 1885 an 8-in. gun was mounted at Fort Lafayette. It consists of four lengths of $\frac{5}{8}$ in. wrought-iron tubing rivetted together and lined with $\frac{1}{8}$-in. seamless brass tubing. The barrel is supported on a braced truss. The breech is closed by a door opening inwards to the side of the valve, and the whole revolves round two trunnions projecting from a cast-iron breech-piece. The trunnions are carried on a cast-iron standard resting on a chassis resembling that in general use for heavy guns. Two cylinders are carried on the chassis; one operates the traversing gear, the other the elevating gear. The hand lever of the firing valve is so placed that No. 1 of the gun can both aim and fire it; the elevating gear is also under his control, and the pressure gauge under his observation. In short, the entire mechanism is under the direct control of one man.

FIG. 114. ZALINSKI'S DYNAMITE GUN.

The air reservoir has a capacity of 137 cubic feet, and is composed of wrought-iron tubes about 1 ft. in diameter, which, in this experimental piece, rested on the chassis. When, however, these guns are mounted in emplacement the air reservoir is placed separately and thoroughly protected. The gun and carriage are also hidden in a pit, the rear half of which can generally be protected by a splinter-proof covering, a traversing range of 180 deg. being sufficient in most positions.

The 8-in. gun has sent shells containing 60 lb. of explosive to ranges of 2¼ miles, and 100-lb. shells up to 3000 yards; 10½ in., 12½ in., and 15 in. guns have been manufactured, and they are intended to throw shells containing charges of 200 lb., 400 lb., and 600 lb. respectively, to ranges approaching two miles, with pressures not exceeding 1000 lb. Fig. 114 shows a 15-in. gun of recent manufacture. The entire arrangement rotates round a fixed vertical cone inside which the air connections are formed to the pipes leading to the reservoir. These guns are made of bronze, and are 40 ft. long, but Captain Zalinski states that the bore can be reduced in length if necessary. The thickness of the tubes forming the bore need not exceed ½ in., but it is generally somewhat thicker in order to obtain rigidity. When weight is important, the tubes can be very lightly constructed, especially if fixed at a constant angle, which can be done in a torpedo boat. When several of these guns are employed in battery (as in the United States war vessel under construction, which carries three guns), a large central air reservoir can be provided in addition to the one serving each gun. The central reservoir can be kept at nearly double the normal pressure, a supply cock provided to each gun reservoir being so constructed that it opens automatically when the firing valve closes. Thus, the rapidity of fire is governed by the speed with which the projectiles can be inserted in the bore, and nearly one round per minute has been obtained.

The Ammunition.—As the present pattern gun is a smooth bore, and the maximum pressure applied small, the shell has thin walls, and a wooden tail like a rocket stick provided with spiral vanes serves to steady it during flight.

The shell is charged with an inner core of dynamite or similar high explosive, surrounded by asbestos paper, and this by an annular charge of nitro-gelatine separated from the shell wall by asbestos. The front of the shell is filled with camphorated nitro-gelatine and a pad of elastic material. The shell usually carries three distinct voltaic (silver chloride) batteries, two of which are wet and one dry. The former come into action when the shell strikes some hard object that collapses the front, thus closing the electric circuit. If the shell fall into water, and the front be not collapsed, the dry battery explodes as soon as the water has wetted it. It is stated that this action can be delayed as desired, so as to insure any suitable submersion before the shell is exploded as a torpedo. A somewhat complicated arrangement is provided for automatically preventing the premature explosion of a shell when in the bore of a gun; but premature explosion just outside the gun is not prevented by it, and this, if it occcurred, would be nearly as disastrous. Electric and percussion fuzes are also fired by the automatic withdrawal of the tail, which is arranged to unscrew from the projectile when

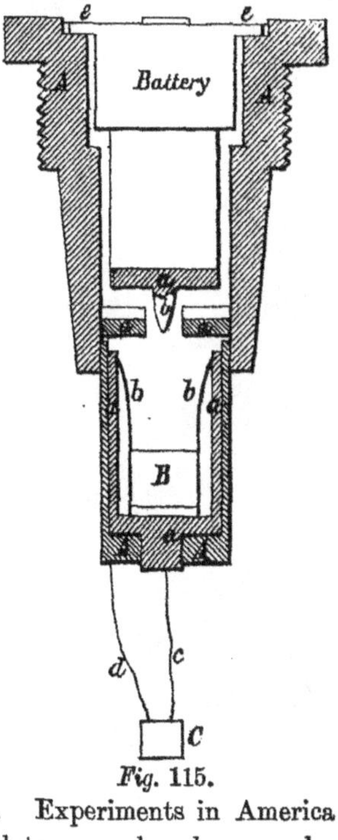

Fig. 115.

it falls into the water. Experiments in America indicate that shells fired against armour plates cause less damage when exploded by percussion fuzes than when exploded by electric fuzes placed in the rear of the projectile. A small chloride of silver battery is carried in case A, Fig. 115. B is a metal plunger in the vulcanite cylinder *a a*; the springs *b b* are connected with one pole of the fuze *c*, the other going to the metal case A. When the gun is fired the inertia of the battery box causes the ears *e e* to be shorn off, and the box takes a position such that B can close the circuit when it moves towards the battery. This occurs as soon as the shell strikes an object. Captain Zalinski has perfected numerous modifications, and has lately patented some of them in this country (*vide* Patent 8995, June 19, 1888, on which date he also took

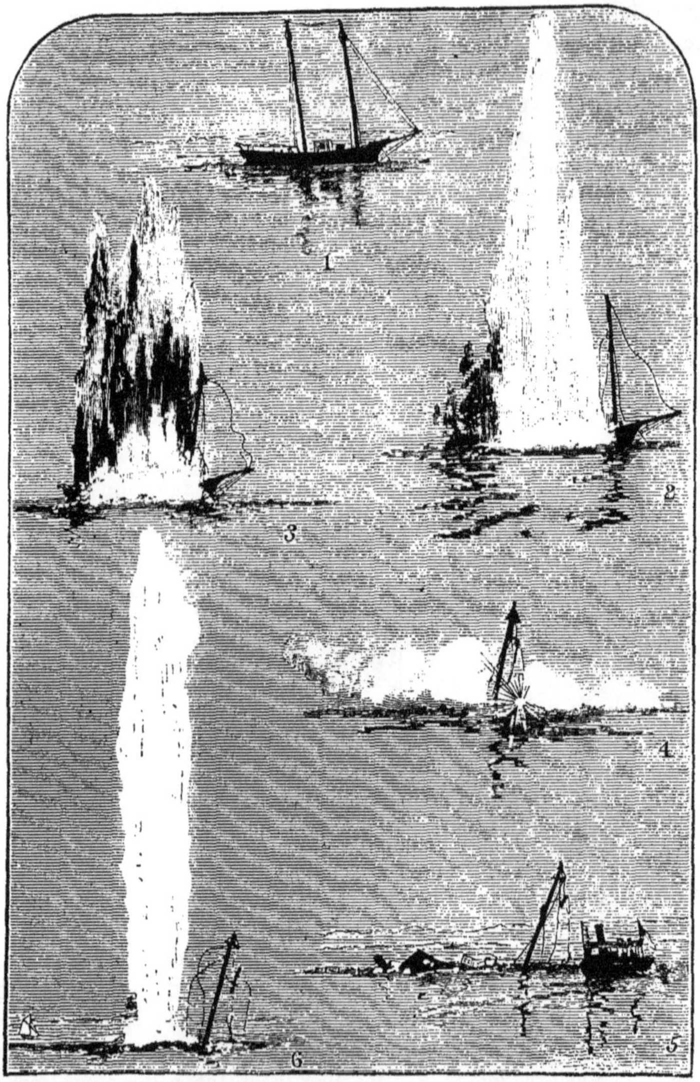

Fig. 116.

out a more complete specification in the United States, No. 384,662). Those who wish to examine the matter minutely can purchase those documents for a few pence, but it would occupy too much space to reproduce them on these pages.

Among other things Captain Zalinski describes a magneto-electric arrangement, to be carried by a projectile, and to be actuated by sudden motions caused by its own inertia, and by changes in the velocity of the projectile. Many of the details are applicable to powder guns.

Accuracy of Fire.—The accuracy obtainable from this artillery is remarkable. In June, 1886, at a trial before the United States Naval Board, "four out of five shells landed in essentially the same spot at a range of 1613 yards," and the fifth "went about seven yards beyond." On the 20th September, 1887, the schooner Silliman was destroyed at a range of 1864 yards (see 1 to 6, Fig. 116). After two sighting shots with blind shell, the third was loaded with 55 lb. of nitro-glycerine, and severely damaged the target vessel (see 2). The second shot destroyed her (see 3). The next struck the wreckage and exploded on the surface (see 4). The last shot exploded at a small submersion (see 6).

Accuracy from a fixed platform on land is therefore established, and on a floating platform the inaccuracy caused by movement is much less than with powder guns. Thus, with 15 min. error in elevation an error of 230 yards at one mile range would be occasioned in an 8-in. rifled gun, whereas an error of only 15 yards would be met with in the air gun.

Comparison with Locomotive Torpedoes.—Torpedo artillery—as developed in the air gun—compares favourably with other methods of carrying a torpedo to any given object of attack.

Compared with the Brennan torpedo:—

1. It is not stopped by booms or netting.
2. Its speed is fourteen times as great.
3. It can discharge torpedoes at the rate of about one per minute.
4. It has no long life artery of wires exposed to injury.
5. Good practice can be made when the object attacked is enveloped in smoke, a mast being all sufficient to aim at.
6. It can be used in thick weather, fog, and darkness (some portion of the object being visible) much better than an arm which must be kept in sight and guided during the whole run.
7. It can be used effectively at short ranges from a rapidly moving platform, such as a man-of-war or a torpedo boat.

In these seven particulars the Brennan torpedo is distinctly inferior, and its superiority on any other point has not been suggested. Torpedo artillery has been developed by Zalinski to a high state of perfection, and the time has arrived to stop any further expenditure on any form of controlled locomotive torpedoes; for no system of the kind can compete with one that hurls large torpedoes with remarkable accuracy and considerable range at the rate of nearly one per minute per gun.

When used at long range, a fixed platform and a good range finder are essential for producing the best results. We possess the range finder, and we ought to have the gun.

There is but little difficulty in so locating and directing torpedo artillery that the shells shall not damage the mine defences, but locomotive torpedoes cannot always be employed over waters mined with electro-contact and automatic arrangements which are near to the surface at low water, especially in situations where there is a considerable tidal range.

Torpedo guns are allied to the artillery of the defence, and their sphere of action is similar. They should be manned by gunners and be directed by artillery officers.

But submarine mines should never be replaced by torpedo artillery or by locomotive torpedoes; for, however perfect the latter may be, they do not possess the blocking effect of hidden and unknown mines.

FINIS.

INDEX.

ERRATA.

Page 38, formula over Fig. 19, for $\frac{P\,D}{9\,l}$ read $\frac{P\,D}{9\,I}$.

" 63, line 5 }
" 64, line 27 } Table XX., page 60, is referred to.
" 67, line 6 }

" 87, Table XXVII. should come at foot of page.

" 150, line 4, for "is situated" read "can be situated."

" 232, line 11, for "efficiency" read "inefficiency."

www.ingramcontent.com/pod-product-compliance
Lightning Source LLC
Chambersburg PA
CBHW030716250326
R18027900003B/R180279PG41599CBX00009B/3
9781845740269